D0977028

GOD'S UNIVERSE

God's
Universe

OWEN GINGERICH

THE BELKNAP PRESS OF
HARVARD UNIVERSITY PRESS
Cambridge, Massachusetts
London, England
2006

The excerpt in Chap. 2 from "The Great Explosion,"
by Robinson Jeffers, appeared in
Tim Hunt, ed., *The Collected Poetry of Robinson Jeffers,*
vol. 3. Copyright © 1954, 1963 by Garth and
Donnan Jeffers for poem.

The excerpt in Chap. 2 from G. Gamow, *My World Line,*
is used by permission, Estate of George Gamow.

Library of Congress Cataloging-in-Publication Data

Gingerich, Owen.
God's universe / Owen Gingerich.
p. cm.
Includes bibliographical references and index.
ISBN-13: 978-0-674-02370-3 (alk. paper)
ISBN-10: 0-674-02370-6 (alk. paper)
1. Religion and science.   I. Title.

BL240.3.G56   2006
215—dc22   2006043502

Contents

FOREWORD

Peter J. Gomes

For the thirty-odd years of my service to Harvard it has been my pleasure to serve as trustee of the William Belden Noble Lectures. This endowment was established by Nannie Yulee Noble in the closing years of the nineteenth century, in support of a series of public university lectures having to do with the Christian religion and the issues of the day; and since then this lectureship has engaged some of the most significant thinkers in the world, often clergy and theologians, on topics of pressing importance. Over the years, however, some of the most significant lectures have been given by lay experts in their respective fields, of whom the most conspicuous—in the first decade of the lectureship—was former president Theodore Roosevelt. In more recent times the lectures have been given by such political figures as Senators Eugene McCarthy and Mark Hatfield; Alan Paton, author and moral citizen; psychiatrist Dr. Armand Nicholi, who lectured on Freud and C. S. Lewis; scientist Dr. Francis S. Collins, director, National Human Genome Research Institute; and television commentator Dr. Timothy Johnson. It was not customary to invite a member of the Harvard faculty to give the Noble Lectures—Professor Harvey Cox having been an exception some years ago—but in extending an invitation to Professor Owen Gingerich to lecture in 2005, I knew that we would be treated to a serious and engaging presentation on one of the most timely topics of the day, that of the relationship between religion and science. The happy results follow.

For three evenings, Professor Gingerich spoke to a large audience of students, faculty, and community members in Harvard's Memorial Church; he is well known at Harvard, having offered many highly regarded courses during his long tenure as professor of astronomy and of the history of science. His reputation as a scholar of the life and work of Copernicus, and his ability to engage the intellectually curious layperson in the substantial matters of astronomy, meant that his audience would take him seriously as one of the university's leading scientific figures. It was less well known that Professor Gingerich was also a devout Mennonite, for whom faith and science had never been at odds. In the heated debates about creationism, intelligent design, and the alleged cultural clash between science and religion, Gingerich, it was widely assumed, would have something interesting and provocative to say.

In the translation of his Noble Lectures into this book, nothing of the elegant encounter with conflicting ideas has been lost, and it is a great good fortune that many more people are now able to gain access to Gingerich's thoughtful discourse. Drawing upon wide experience and deep scholarship, and armed with often disarming understatement where others frequently hurl absolutes, Gingerich helps to define issues in such a clear way that we emerge from the discussion with the ultimate academic compliment: Why didn't I think of that?

For some, the intellectual modesty with which Gingerich approaches the subject may seem too modest, for neither "godless" scientists nor the devout creationists will find here sufficient fuel for their broadsides. Gingerich, however, is not diffident, in the sense of not daring to say anything, but humble in the face of the enormity of his task. He is modest in the sense that Albert Einstein, whom he quotes frequently, was modest, for it was Einstein who said, "The sense experiences are the given subject matter [of science]. But the theory that shall interpret them is man-made . . . never completely final, always subject to question and doubt." In speaking of the constraints of science, a concept familiar to genuine scientists but alien to those who invoke science for their own purposes, Gingerich observes in his prologue: "Science works within a constrained framework in creating its brilliant picture of nature. But reality goes much deeper than this. Scientists work with *physics,* but (perhaps unwittingly) they have a broader system of beliefs, *metaphysics,* a term that literally means 'beyond physics.'" Gingerich discusses what he calls a "scientific tapestry of the physical world," but he is quick to note that scientists "also wrestle with the metaphysical framework within which the universe can be understood." He regards intelligent design as "misguided when presented as an alternative to the naturalistic explanations offered by science, which do not explicitly require the hand of God." He goes on to say, "This does not mean that the universe is actually godless, just that science within its own framework has no other way of working."

Listening to Gingerich as a man of faith and of science makes it possible to realize that these two great ways of thinking need not necessarily be pitched at odds with one another, unless one side is greedy, jealous, or uninformed about the other. This book, like the lectures that gave it birth, offers conviction and clarity about some of the most vexed issues of our day, and where there are no clear answers, Gingerich does not presume to offer any. His closing chapter is styled, "Questions without Answers," which title suggests that in this pilgrimage of heart and mind the journey is still on, the destination not yet reached. As Dante was fortunate to have Virgil as his guide, we are fortunate in having Owen Gingerich as ours.

GOD'S UNIVERSE

PROLOGUE AND
PILGRIMAGE

When that April with his showers soote
The drought of March hath pierced to the roote . . .
When Zephirus eek with his sweete breath
Inspired hath in every holt and heath
The tender croppes, and the yonge sonne
Hath in the Ram his halve-course yronne . . .
Then longen folk to goon on pilgrimages . . .

— Geoffrey Chaucer, Prologue, *Canterbury Tales*

I grew up in small towns in middle America. On a summer night when I was five, the temperature in the house at sunset still stood at over a hundred degrees, so my mother took cots outside for sleeping. I looked up at the darkening sky and asked, "Mommy, what are those?" to which she replied, "Those are stars—you've often seen them!" I am reported to have responded, "But I never knew they stayed out all night!"

And that was the beginning of my love affair with the stars. My father encouraged me by helping build a simple telescope from a mailing tube and surplus lenses from a local optometrist. We could see the rings of Saturn, and I was mightily impressed when he devised a camera to record with the telescope a partial eclipse of the sun. But the earth, too, held intriguing curiosities for an inquisitive lad, and I soon became an eager collector of rocks and fossils. My parents and the community nurtured my interest in astronomy. Geology and evolution may have been regarded with some suspicion, but for the most part they were simply passed over in silence.

We were a religious family. My father's four great-grandfathers were all Amish ministers, and we stayed within an Anabaptist tradition that emphasized believer's (i.e., adult) baptism and the pacifist ideals of the Sermon on the Mount.

In 1947, because my parents moved to Indiana, I enrolled in Goshen College, a Mennonite school whose motto was "Culture for Service." And that motto epitomized my dilemma. While an undergraduate, I had the marvelous opportunity to go to Harvard College Observatory as summer assistant for Harlow Shapley, then the most famous astronomer in America. At Goshen I was a chemistry major, and I knew that chemistry provided many opportunities for service to humankind. Meanwhile the stars and Harvard beckoned, but of what practical use could astronomy be? Here my mathematics professor came to the rescue. "If you feel a calling to pursue astronomy," he counseled, "you should go for it. We can't let the atheists take over any field."

And so it was that I entered graduate school in astrophysics. But my interests were never limited to astronomy. I audited courses in physics and geology and went along with the paleontology students on their fossil-gathering field trips. I was introduced to the history of science when I became a teaching fellow in one of the distribution courses in natural science that adopted President James B. Conant's case study approach. This method employed historical episodes to illustrate the interplay of observation, experiment, and the development of scientific understanding. Thanks to the philosophical readings assigned in that course, I became increasingly curious about the fundamental way that science worked, its claims to truth, and the relation of those claims to religious belief.

Eventually I finished my astrophysics dissertation and became a scientist at the Smithsonian Astrophysical Observatory (which had meanwhile moved to Cambridge)

3

and later a professor of both astronomy and history of science at Harvard. For over a decade I used the powerful electronic computers, new at the time, to trace the way light flows through the outer layers of stars; in this way I was able to analyze the differences between dwarf and giant stars and distinguish between stars with more or less of the heavier elements. Some of these calculations were the first to take into account rocket and satellite observations of the sun and other stars.

I have often remarked that if you want to know how science works, there is a great advantage in actually working as a scientist, trying to tease out the structure of nature from among the ambiguities of the observations made at the cutting edge of the field. But a complementary approach is to examine some historical cases in depth, because the passage of time affords a rich and helpful perspective. It is particularly fascinating to realize that today we are still facing some of the same issues concerning the interface of science and religion that Copernicus, Kepler, and Galileo grappled with—notably, the role of scriptural literalism. An attentive reader of the lectures that follow will notice that both avenues—scientific research and historical research—inform their content.

In the 1980s two opportunities moved me more visibly into the science-religion dialogue. First, I was invited to inaugurate a new series at the University of Pennsylvania, the Dwight Lectures in Christian Thought, for which I prepared an address entitled "Let There Be Light: Modern Cosmogony and Biblical Creation." This essay, which I sometimes referred to as my pro-Christian/anti-Creationism lecture, was subsequently published in Roland Frye's anthology *Is God a Creationist?* and in an updated version in Timothy Ferris's *World Treasury of Physics, Astronomy, and Mathematics.*

Second, largely as response to the great success of Carl Sagan's *Cosmos* television series, which offered a conspicuously materialist approach to the universe, the American Scientific Affiliation (an organization of Christians who take both science and the Bible seriously), under the leadership of its executive director, Robert Herrmann, urged me to plan a television science series with a different philosophical spin. Geoff Haines-Stiles, who had been the producer of *Cosmos,* worked with me to develop treatments for six episodes of *Space, Time, and God.* Although we never received the funding to make such a major series possible, the exercise helped focus my thoughts on the basic issues where science and religion intersect. How, you may well wonder, would such a series have differed from the more typical media approach to science?

To begin with, rather than dwell on historical conflicts between science and religion, the program would have pointed out that the Judeo-Christian philosophical framework has proved to be a particularly fertile ground for the rise of modern science. Next, rather than declare that humankind is a marginal and merely fleeting occurrence in a universe so vast and ancient, the series would have explained that we would not be found in a smaller, young universe because there would not have been time in it for the slow "cooking" of the elements required for life. Then such a series would have illustrated how remarkably hospitable and how suited our universe is to the development of intelligent life. And finally, the program would have spoken to the nature of science itself. Science is not simply a collection of facts; it is a grand tapestry, woven together from facts and the hypotheses that unite these facts in an encompassing pattern of explanation. As Einstein put it, "the sense experiences are the given subject matter. But the theory that shall interpret them is man-made . . . never completely final, always subject to question and doubt."[1]

Science works within a constrained framework in creating its brilliant picture of nature. But reality goes much deeper than this. Scientists work with *physics,* but (perhaps unwittingly) they also have a broader system of beliefs, *metaphysics,* a term that literally means "beyond physics." The lectures that follow present a scientific tapestry of the physical world, but they also wrestle with the metaphysical framework within which the universe can be understood. They argue that the universe has been created with intention and purpose, and that this belief does not interfere with the scientific enterprise. But I further contend that the current political movement popularly known as Intelligent Design is misguided when presented as an alternative to the naturalistic explanations offered by science, which do not explicitly require the hand of God. This does not mean that the universe is actually godless, just that science within its own framework has no other way of working.

Figuratively speaking, life is a pilgrimage, a journey of learning: learning from teachers and testing, from books and the media, from adventures and contemplation. It is an unfinished quest for understanding, a work in progress. Likewise, the three William Belden Noble Lectures delivered in November of 2005 and presented in this volume are a work in progress, a status report, as it were. Assembled in these lectures are reflections and insights gathered along the pilgrimage trail, some that I came upon decades ago, others as fresh as yesterday. I hope that the foregoing account of my personal pilgrimage will supply the reader with a basic roadmap to the lectures that follow.

The Andromeda Galaxy—about two million light-years away in the constellation Andromeda—an island universe comparable to our own Milky Way Galaxy. The peppering of points filling the scene consists of foreground stars belonging to the Milky Way. Photographed with a four-inch telescope and CCD camera by Dennis di Cicco, October 2003.

1

Is Mediocrity
a Good Idea?

So vast, without any question, is the Divine Handiwork
of the Almighty Creator!

—Copernicus, *De revolutionibus*, 1543

*C*uriously, the soaring passage from Copernicus' *De revolutionibus (Revolutions)* found at the end of his spirited defense of a sun-centered cosmology, was specifically censored by the Vatican when his *Revolutions* was placed on the *Index of Prohibited Books*. Obedient Catholics pasted strips of paper over the offending text, or inked it out. Galileo, who aimed to persuade the authorities that he was on good behavior, dutifully struck out the sentence, but with a fine line, so that Copernicus' pious exclamation could still be read: "So vast, without any question, is the Divine Handiwork of the Almighty Creator!"

A century earlier the advent of printing had swept over Europe like a tsunami, bringing a wave of fresh ideas to a wide range of readers. In England John Wycliffe dared to transpose the Bible into the vernacular, and forty-four years after his death the pope ordered Wycliffe's bones exhumed and burned. Less than a century later, Martin Luther would translate the Bible into German, and in 1514 he had had the audacity to print the Psalms without the traditional marginal references to the interpretations of the Church Fathers. "Scripture alone!" became the slogan, and even Rome was driven to a defensive scriptural literalism.

Did Psalm 104 not say, "The Lord God laid the foundation of the earth, that it not be moved forever"? Surely a sun-centered cosmology, with the earth spinning daily on its axis and whirling annually around the sun, was ruled out by sacred scriptures. The only way to allow Copernicus' unorthodox cosmology to be read was to brand it a mere hypothesis, but Copernicus' pious exclamation seemed to claim that his was the way God actually made the cosmos, so that passage had to be excised. Nor were the Catholics alone in their fear that the heliocentric cosmology would make the Bible false. A Protestant clergyman added an anonymous foreword to Copernicus' book, saying, "These hypotheses need not be true nor even probable," and as late as the 1660s the

Dutch Reformed Church nearly split over the issue of Copernicanism.[1]

Fast-forward to the twenty-first century. Millions of people who would never dream of declaring that pi, the ratio of the circumference of a circle to its diameter, is exactly three because of the account in 1 Kings 7:23 (of a round metal casting ten cubits in diameter and thirty cubits in circumference) nevertheless believe that the world was created in essentially its present form only a few thousand years ago. Folks who take in stride the modern technology of cell phones, laser scanners, airplanes, and atomic bombs nevertheless show reluctance to accept the implications of the science that lies behind these awesome inventions of the past century. The paradox is profound and deserves sober reflection. An underlying theme of the investigation that follows is how to take both science and the Bible seriously.

To set the stage for these reflections, let us consider a simple question posed by Sir John Polkinghorne: "Why is the water in the teakettle boiling?"[2]

We can answer: "The water is boiling because the heat from the fire raises the temperature of the water until the molecules move faster and faster so that some escape from the surface and become a gas." But we can also answer that the water in the teakettle is boiling because we want some tea. The first answer illustrates what

Aristotle called an efficient cause, an explanation of how the phenomenon takes place, while the second answer, "Because we want some tea," is a final cause, the reason the phenomenon takes place. One aspect of the scientific revolution of the seventeenth century was that it turned away from the final causes so central to the Aristotelian worldview and concentrated on efficient causes, the how of the phenomena.

To me, belief in a final cause, a Creator-God, gives a coherent understanding of why the universe seems so congenially designed for the existence of intelligent, self-reflective life. It would take only small changes in numerous physical constants to render the universe uninhabitable. Somehow, in the words of Freeman Dyson, this is a universe that knew we were coming.[3] I do not claim that these considerations are proof for the existence of a Creator; I claim only that to me, the universe makes more sense with this understanding.

At the same time, as a scientist I am interested in the how as well as the why of the universe. I want a coherent picture of how *Homo sapiens* came into being, how our DNA can be so wondrously related to that of all forms of life, how the atoms emerged. The answers to these questions would all fall within the realm of efficient causes, and it is the language of that realm which will echo throughout these lectures. So I come to you as a professional scientist and a historian of science, but also as an amateur theologian. And I begin with a question that may well pique your curiosity: Is mediocrity a good idea?

Surely at Harvard (where these lectures originated and where all the children are above average), the answer is a resounding NO! We could not for a moment entertain the idea that mediocrity is a good idea. When, a few years ago, Garrison Keillor, the legendary storyteller of National Public Radio's *Prairie Home Companion,* came as the Harvard/Radcliffe Phi Beta Kappa orator, his tongue-in-cheek monologue was a defense of mediocrity. He suggested that it would be good therapy for a professionally driven caste always clambering to the top of the heap. The audience was delighted to be teased with this sophisticated, in-your-face humor, but of course no one believed a word of it.

Yet I would like to consider the notion of mediocrity seriously, because in the latter half of the twentieth century mediocrity gained considerable cachet as a scientific tool, under the heading "the Copernican principle": the precept that we human beings cannot flaunt a unique or special identity. The idea is that we will make scientific progress if we consider that everything we see around us is commonplace in the universe, that we are average beings in a run-of-the-mill planetary system in an average galaxy probably populated by scores of other mediocrities.

A central figure in the scientific quest to understand our place in the cosmos was the sixteenth-century astronomer Nicolaus Copernicus. A churchman affiliated with the Frauenburg Cathedral in the northernmost Catholic diocese of Poland, Copernicus dreamed of a "theory pleasing to the mind" and worked out the idea of a heliocentric cosmology. If the sun, rather than the earth, was at the center of the universe, the apparent, complex motions of the planets could be explained more simply. Despite Copernicus' proposal, the time-honored and seemingly sensible conception of an earth-centered universe did not die quickly. Nevertheless, the new heliocentric scheme gradually took root as an actual physical description of the world.

Copernicus, by transforming the earth from a unique, central place in the cosmos to just one planet among many, essentially invented the concept of the solar system. In the steps that followed a century later, the sun, by then recognized as a star, became only one of many. In the twentieth century it has become increasingly popular to refer to a "Copernican principle," namely: We should not consider ourselves to be on a special planet circling round a special star that has a special place in a special galaxy. With respect to the cosmos we should not be considered special creatures, even though we clearly are with respect to life on earth. In full dress, this is the principle of mediocrity, and Copernicus would have been shocked to find his name associated with it.

Copernicus took the first step, in showing us that we were not the center of the universe after all, and Charles Darwin took another, in proposing a mechanism that could in theory populate a plenitude of planets, though as a thoroughgoing Victorian he was deeply convinced that Englishmen were at the pinnacle of the evolutionary tree. In any event, this was a far cry from the Psalmist's exclamation, "What is man that thou art mindful of him? For thou hast made him a little lower than the angels."

As I was beginning to formulate these lectures, there arose a considerable flap at the Smithsonian's National Museum of Natural History in Washington, D.C., because the museum had agreed to show a film entitled *The Privileged Planet*. I had a minor walk-on role in the controversy because I had written a dust jacket blurb for the book of the same name.[4] Without my knowledge, the sponsors of the film had used my rather cautious endorsement, where I had said in part, correctly, "This thoughtful, delightfully contrarian book will rile up those who believe the 'Copernican Principle' is an essential philosophical component of modern science." Someone from the museum staff had previewed the film and had found no problem with its science; however, the film was sponsored by the Discovery Institute in Seattle, a think tank well known as a principal proponent of the so-called Intelligent Design movement, and very quickly critics raised the alarm that the showing of the film *The Privileged Planet* at the Smithsonian Museum would somehow constitute an endorsement of Intelligent Design. I suppose that few of the critics actually saw the film, for it contains no explicit mention of Intelligent Design.

It did, however, contain implicit criticism of the Copernican principle, for the film argued that the earth is indeed a very special place, something that we would all intuitively agree with, since it is, after all, our home. But the film carried its assertions to a cosmic level, in proclaiming how very special, how unique, in fact, our planet's location and circumstances are. The implicit message of the film was that we, all members of the species *Homo sapiens,* have been endowed with a highly unusual environment, not only conducive to our existence here, but also remarkably well suited as a vantage point from which to investigate the cosmos itself. Who can fail to be thrilled by the idea that we have inherited a place uniquely situated for surveying the universe? The answer is, those who have adopted the Copernican principle as a rule of science fail to be thrilled, along with those who, by extension, feel that any implied criticism of that principle of mediocrity is an attack on science itself.

But does the so-called Copernican principle gain us any leverage scientifically? What does a careful and critical examination of the use (or nonuse) of this principle actually show concerning the development of our scientific understanding of the universe? I have been able to think of four, and only four, episodes in the history of science where the Copernican principle *could* have helped enhance our scientific knowledge, and I beg your indulgence for this brief technical excursus. Let us count how often it was actually used.

The first concerns Copernicus himself and his radical new concept of a sun-centered cosmos. In his time, in the sixteenth century, fixing the sun at the center and throwing the earth into dizzying motion seemed completely ludicrous, a violation of common sense. What Copernicus discovered was that if you place all the planets, including the earth, into orbit around the sun, the planets are automatically arranged in order of their periods of revolution. As Copernicus himself boasted, "In no other way do we find such a sure harmonious connection between the size of the orbit and period of the planet."[5] It was such a beautiful unifying principle that there was no turning back. Common sense be damned! If placing the planets in orbit around the sun produced such an elegant and orderly progression, just ignore the ridiculous and theologically dangerous spinning of the earth.

Copernicus, in transforming Ptolemy's ancient geocentric arrangement into the heliocentric solar system, supposed that the earth moved in a circular orbit and *at a constant speed.* That made the earth unique among the planets, for all the rest of them essentially moved in their circles *at variable speeds.* Now, if Copernicus had applied the Copernican principle, he would have used the same mechanism for all the planets, but he didn't. He treated the earth as a special case; and unbeknownst to him, his unique arrangement for the earth led directly to serious errors in predicting the positions of Mars. It remained for Johannes Kepler, a few generations later, to mend this Achilles' heel in the original Copernican system.[6] Still, the Copernican principle was not explicitly invoked here. Kepler's motivation, grounded strictly in physics, derived from the idea that if the sun supplied the motive power for the planetary revolutions, then the earth should orbit faster when it was closer to the sun. Let's not get bogged down in details, fascinating as they are to specialists. Suffice it to say that in this critical case,

Copernicus had failed to recognize the potential significance of a Copernican principle.

My second example concerns the size of the solar system. With breathtaking boldness, the ancient Greek mathematical philosophers undertook to figure out how far away the sun was. Actually, they determined the size of the earth and the distance to the moon with respectable accuracy, but their admirable attempts to establish the distance to the sun led to a spurious answer, a figure twenty times too small. With the invention of the telescope early in the seventeenth century, astronomers gained a fresh way to tackle the problem, because for the first time they could see and therefore measure the angular diameters of the planets. The method is essentially one we intuitively use to gauge how far away persons in the distance are. We automatically assume that everyone is about the same size, and the smaller they appear, the farther away we deduce them to be. From the time of Copernicus on, the relative distances in the solar system were known, so measuring how far away any planet was from the earth (in miles, for example) would automatically establish the scale of the system and therefore the distance to the sun in miles. If one of the seventeenth-century astronomers had, assuming that the earth was about the same size as the rest, applied the Copernican principle to the planets and had averaged the deduced distances to Venus, Mars, and Jupiter on the assumption that they were like the earth, he could have come up with a vastly improved scale for the solar system, not wide of the mark.[7] Score zero for the Copernican principle, though, because it scarcely occurred to anyone to apply it in this case. The one person who came close, the Dutch polymath Christiaan Huygens, was astute enough to realize that the planets were *not* all the same, but he assumed that the earth should fit harmoniously in size between Venus and the appreciably smaller Mars. Despite the fact that his assumption was not perfect, he succeeded in proposing a much improved scale, though at the time no one knew whether or not to believe it.[8]

The third example was more influential. When astronomers finally realized, in the seventeenth century, that stars were distant suns, several of them independently thought of a way to gauge their distance from the solar system. Basic to their scheme was the assumption that the sun was just an average star, and that some stars appeared brighter simply because they were closer to us. This Copernican principle of mediocrity was essential to the outcome arrived at by the early theorists—which was of course flawed because stars are not in fact like peas in a pod, but at least James Gregory, Isaac Newton, and Christiaan Huygens all came up with figures for the distance to the star Sirius that were in the ballpark; thus, by 1700 astronomers had a fair idea how far away from us the stars are.[9]

The fourth episode, from the twentieth century, is another missed opportunity, however. When I came to Harvard as a graduate student in 1951, there was a scary skeleton in astronomy's closet. It was well known that the distant galaxies were rushing away from us, and the farther away they were, the faster they raced in this headlong expansion of the universe. From their speeds and distances we could calculate the time that must have elapsed since a great primeval explosion. The problem was that the length of time, about two billion years, seemed too short, for geologists were sure that the earth contained rocks that were older than that, and it is hard to imagine a universe younger than its parts. Then there was another puzzle, seemingly unrelated. The disk of our Milky Way Galaxy is surrounded by a halo of a couple of hundred globular clusters, huge balls of hundreds of thousands of stars, and so is our neighboring island universe, the great Andromeda Galaxy (frontispiece to this chapter). It did seem odd that as a rule Andromeda's globular clusters were systematically fainter than ours— so odd, in fact, that Harlow Shapley, who was then director of the observatory at Harvard, when he taught a graduate course in cosmology in the spring of 1952, assigned this puzzle as a term project to one of the graduate students. Now, if Shapley and his student had applied the Copernican principle to those globular clusters and had reasoned that the Milky Way's clusters could not be brighter than the others because that would put us in a special place, curiously endowed with brighter-than-average globular clusters, then they would have been forced to conclude that something was radically wrong with the accepted distance from us to the Andromeda Galaxy. If that galaxy were twice as far away, for example, its globular clusters would naturally appear fainter, and that could account for the seeming disparity. And if the Andromeda Galaxy was two million light-years away instead of one million, say, the whole scale of the universe would have to recalibrated, and consequently its age.

Alas, neither Shapley nor his student thought of this. And few have had the temerity to challenge the foundations of astronomy by applying the principle of mediocrity to the problem. But that very summer, in 1952, following an entirely different course of observation and investigation, the astronomer Walter Baade, working with the giant telescopes in southern California, realized that the scale of the universe had been incorrectly gauged. He thus cut that Gordian knot with a single stroke. It would have been a brilliant opportunity for the Copernican principle, but I suspect that astronomers would have written off such a basis for a proposed solution as too wildly speculative.

So, in the four possible applications, the Copernican principle has been implicitly invoked only once.

Now, given the spotty historical record concerning the actual usefulness of the Copernican principle of mediocrity, you may well wonder why it is so popular today and why many astronomers can get downright feisty when you challenge it. So another brief excursus is in order. In 1796 the French astronomer and mathematician Pierre-Simon Laplace proposed that stars and planets condensed out of swirling disks of gas and dust. This theory nicely explained why the orbits of all the planets lie pretty much in the same plane, and why they revolve around the sun in the same direction. It made sense that if stars condensed out of a nebula, there should then also be a multitude of other planetary systems around the distant stars, too far away to be readily detected.

Laplace's nebular hypothesis held sway for nearly a century, until serious mathematical objections were raised against it. These had to do with what is called conservation of angular momentum, a phenomenon famously demonstrated by skaters when they go into a spin by pulling in their outstretched arms and leg. The sun is effectively doing the same thing by pulling in the surrounding gas and dust. The problem is that the sun is not spinning fast enough. Because the sun has the lion's share of mass in the solar system, it would be expected to possess most of the angular momentum, but the sun would have to spin much faster just to match the angular momentum of the planet Jupiter, and still faster to have its proportionate share. Then an alternative scenario was proposed. Perhaps planetary systems are formed by near-collisions of stars, cataclysmic events in which material torn out of these traumatized masses goes into orbit and condenses into planets. But such an event would be exceedingly rare. If I represent the sun by a marble about an inch in diameter, the next-closest star in this scale model would be another marble located about six hundred miles away. The chances that these stars might collide would be almost vanishingly small, and our earth would truly be a very privileged planet in a galaxy with only a tiny number of planetary systems. In such a cosmos, habitable environments would be rare indeed, and other environments where intelligent life could exist might be limited to the planet Mars.

This all changed in the 1950s. By that time the collision hypothesis had in turn met with its own powerful criticisms, and the nebular hypothesis had been rehabilitated because ways had been proposed whereby the sun could shed its angular momentum and therefore be spinning at its observed rate of about once a month, without violating the principle of the conservation of angular momentum. Assuming that planets can be formed of the same clouds of gas and dust out of which the stars themselves are born, then we can envision billions of planets within the Milky Way Galaxy, and millions of habitable environments. Close on the heels of this rehabilitation of the nebular hypothesis came the development of radio astronomy, and around the same time a seminal paper written by the late Philip Morrison and his colleague Giuseppe Cocconi, which proposed that a search for extraterrestrial intelligence could be undertaken by looking for evidence of radio signals from remote civilizations.[10]

On several occasions Philip Morrison took part in debates in my class on whether we are alone in the Milky Way. He built his argument in favor of extraterrestrial intelligence around three code words: antiquity, plenitude, and ubiquity. Antiquity: our universe is unimaginably old, something over ten billion years, which allows a long time for nature to experiment with building life forms. Plenitude: there are plenty of habitable places where life might flourish. And ubiquity: everywhere the same laws of physics and chemistry hold true, so if life developed here, it could do the same elsewhere. Antiquity, plenitude, and ubiquity—and tacitly was he not drawing on the Copernican principle, which is deeply entangled with plenitude and ubiquity? How dare we think we are so special, so privileged, that intelligent life would not arise aplenty and everywhere! Hence, mediocrity has become the banner under which the search for intelligent life elsewhere goes on. More an ideology than a scientific law, the Copernican principle provides the bedrock on which SETI, the search for extraterrestrial intelligence, rests.

Indeed, in a certain sense the time is finally ripe for the discovery of extraterrestrial signals. In humankind's long journey from the taming of fire to modern technological prowess, we have not only achieved the ability to radically alter the earth and its atmosphere, but we have attained an understanding of the nature and place of distant stars, along with knowledge of how to send or receive signals across vast distances. A century ago this endeavor would have been quite impossible. Today such communication is well within our ken. As Morrison and Cocconi remarked at the end of their trailblazing paper, "The probability of success is difficult to estimate, but if we never search, the chance of success is zero."[11]

The idea that other intelligent beings might be out there is not particularly new. With the very first inkling that other celestial worlds might resemble our own, imaginative writers began to populate those worlds with other creatures. The seventeenth-century German as-

26

tronomer Johannes Kepler wrote a pioneering science fiction dream in which he described the inhabitants on the moon, though admittedly it was a form of Copernican propaganda to show his public how celestial movements would appear from a moving body different from the earth.[12] His younger contemporary, the chaplain John Wilkins (later bishop of Chester), described in vernacular English an imaginary voyage to the moon.[13] A century later Christiaan Huygens explained why the inhabitants of Saturn would have hemp: "If their Globe is divided like ours, between Sea and Land, as it's evident it is, we have great reason to allow them the Art of Navigation, and not proudly ingross so great, so useful a thing to our selves. And what a troop of other things follow from this allowance? If they have Ships, they must have Sails and Anchors, Ropes, Pullies, and Rudders."[14]

Now in a new millennium, we have learned of many other far-flung celestial worlds that are potential homes for extraterrestrial life. We know as well that there are about two hundred billion stars in our Milky Way Galaxy (more than thirty apiece for every man, woman, and child on our planet), and we know that more than a hundred billion galaxies exist beyond the Milky Way system. We have every reason to believe that planets circle many of these distant stars (even though they are too faint to be examined in any detail with our present instruments), and given the wealth of possibilities, countless habitable environments must be scattered throughout these starry realms. Conservative speculators concede that on some there may be life. Enthusiasts argue that there *must* be life, and some of it will inevitably be intelligent life, including creatures possessing intelligence far beyond our own.

Science would scarcely function without the assumption of the uniformity of nature, a principle closely akin to the Copernican principle of mediocrity, but more universal. To envision a universe with other *habitable* planets seems well within the range of conjecture fostered by this assumption, but to expect that the uniformity of nature will necessitate the existence of *inhabited* planets is undoubtedly a stretch of the principle. Hence the rhetoric from the enthusiasts: it is the Copernican principle that is popularly cited as an argument in favor of the interpretation of a universe teeming with alien intelligent life.

Theologically, the way has been open to consider the possibility of extraterrestrial life ever since the bishop of Paris declared in 1277 that it was heretical to limit to just the earth God's power to create life.[15] This edict placed the Church in an ambiguous and awkward position, for the ethos of its teachings was that mankind occupied the pinnacle of Creation; and surely the stories in Genesis 1 and 2 reach a crescendo with the creation of Adam and Eve. Whether or not God would have chosen to create intelligent life elsewhere was another question, and most churchmen thought not. Today, at least, we have the possibility of putting that question to the test.

In accepting the principle of mediocrity, the *E pluribus unum* approach, we have had the arrogance to assume that everyone else was sufficiently mediocre and uniform to communicate our way, but if a great range of intelligence actually exists out there, our way might be pretty primitive by cosmic standards. This is not to criticize the present search strategy—after all, we must start somewhere if we want any chance to test the hypothesis—but we should be ready to recognize our unwitting anthropocentrism.

Yet anthropocentrism is not necessarily bad or even wrong. It is well to remember that the human brain is by far the most complex physical object known to us in the entire cosmos. Of the roughly thirty-five thousand genes coded by the DNA in the human genome, half are expressed in the brain. There are about a hundred billion neurons in the brain, nerve cells, many with long dendritic extensions intricately interconnected with each other. Each neuron connects with about ten thousand other neurons. While the estimated number of stars in all the galaxies in the universe vastly exceeds the number of grains of sand on all the beaches of the world, the number of synaptic interconnections in a single human brain vastly exceeds the number of stars in our Milky Way: 10^{15} synapses versus about 10^{11} stars.

For an eleven-year-old at rest, roughly half the body's energy supply fuels the brain. (For a newborn it is 74 percent, for an adult 23 percent.)[16] Oxygen (required for slowly "burning" the organic fuel) is carried to the brain by the red blood cells. In these cells the oxygen is loosely bonded to the iron atoms in the middle of the heme complex in the blood's hemoglobin. The oxygen is transferred into the blood through the foliate lung system, intricately branched, where the solubility of oxygen in water and the diameter of the capillaries are fine-tuned for an efficient rate of transfer of the oxygen to the heme. Of all the metallic complexes, iron has just the right bonding strength to allow the capture and subsequent easy release of the oxygen.

Fortunately for us, our atmosphere contains a reasonable supply of oxygen—about 20 percent by number of atoms. This percentage is high enough to sustain fire, but not so high as to allow cataclysmic combustion. In fact, the acceptable oxygen limits for life more complex than single cells are fairly narrow, and the earth's atmosphere, like the little bear's porridge, seems just right. In 1913 the Harvard chemistry professor L. J. Henderson drew attention to this fact in his remarkable book *The Fitness of the Environment,* and Michael Denton has more recently updated and detailed even more such extraordinary circumstances in his book *Nature's Destiny.*[17]

As the human brain develops from infancy, a substantial part is devoted to the control of the organs of speech. No other aspect of human powers differs so significantly from that of the other animals as our ability to communicate by spoken language. The Harvard anthropologist David Pilbeam has remarked that if we could have observed Neanderthals over the millennia, we could hardly have extrapolated to the complex human civilization that eventually arose on earth. In the two hundred thousand years of their existence, the Neanderthals' stone tools remained without improvement, as if frozen in time. In contrast, *Homo sapiens sapiens* reached a point where he began, gradually, to improve his tools. Perhaps the lack of progress among the Neanderthals was owing to their lack of language. Their stasis should certainly give pause to those who believe that the evolution of intelligence is inevitable.

In any event, the evidence at hand is hardly conducive to modesty. Human beings, with their brain capacity, their use of complex language, and their ability with abstract reasoning, clearly represent the pinnacle of life on earth, far outdistancing any rivals, and to say otherwise is to engage in a sort of scholastic fantasy. "What is man that thou are mindful of him?" asks the Psalmist. "For thou hast made him a little lower than the angels and hast crowned him with glory and honor." Yet part of the glory of human creativity and self-consciousness is the ability to ask questions reaching beyond ourselves, about whether the human brain is really the most complex object in the universe or about whether we are alone in the universe—alone in either sense, whether God exists or whether extraterrestrial intelligence exists.

Let us pursue this inquiry further. I have outlined in only the most rudimentary fashion two remarkable conclusions. Each would require an entire book to defend it adequately. The first conclusion is that human beings are astonishingly well constructed within the framework of possibilities and that the cosmic environment in general and the earth's environment in particular are themselves wonderfully congenial to intelligent, self-conscious life. The second conclusion, less well delineated here, is that intelligent, self-conscious life was not necessarily inevitable in our planetary system and, by extension, is not necessarily inevitable elsewhere.

Historically, until the work of Charles Darwin, the first conclusion was explained as evidence for the designing hand of a beneficent Creator, epitomized by William Paley's 1802 book, *Natural Theology; or, Evi-*

dences of the Existence and Attributes of the Deity. Darwin's *On the Origin of Species* (1859) offered a naturalistic alternative: the mysterious arrival of variations followed by natural selection of the fittest varieties gradually left the world full of creatures that were singularly adapted to their environments. Darwin's evolutionary scheme offered a reasonable explanation for the variations in organisms, in both time and place and, in its key reliance on common descent, helped biologists understand the relationships of plants and animals and especially the curious "imperfect" adaptations, such as the web-footed ducks that nest in trees in the Galapagos. As understanding of genetics and then of molecular genetics has increased, the idea of common descent and organismic relationships has become even more firmly ensconced.

The evolutionary picture is one of a zigzag, opportunistic process. Irven DeVore has remarked that if the ancient lungfish, crawling onto the shore, had turned left instead of right, the evolution of land vertebrates would not have yielded our present fauna. Stephen Jay Gould, in describing the strange life forms that lived in the Middle Cambrian ocean, as evidenced by the famous Burgess Shale fossils, emphasized the role of contingency and accident; if the tape of life were rewound and played out again, he declared, the results would have been unpredictably different. Life would "cascade down another pathway," and the chance that the rerun would "contain anything remotely like a human being must be effectively nil."[18] David Pilbeam, in criticizing the implications of the film *Planet of the Apes*, where chimpanzees take over as the intelligent masters after a nuclear disaster has eliminated *Homo sapiens*, says, "If we wipe ourselves out through a nuclear catastrophe, don't expect that evolution will ever again replace humankind with anything like us" (David Pilbeam, personal communication).

These Darwinist scenarios reinforce the conclusion that the evolution of intelligent, self-conscious life elsewhere is by no means assured or even probable. The late dean of evolutionists, Professor Ernst Mayr, felt so strongly about the absurdity of extraterrestrial intelligence that he always declined to come to my class even to debate the issue, but he did provide an essay in which, while agreeing that it is quite conceivable that life could originate elsewhere in the universe, he argued that such a process would "presumably result in living entities that are drastically different from life on earth."[19]

The Darwinian viewpoint has interesting implications for the principle of mediocrity. While insinuating that we are the unplanned outcome of a naturalistic, mechanistic process—that is, essentially a glorious accident—it also leaves open the prospect that we are in fact at the top of the heap, or at the very least we are on a peak so different from the heights perhaps occupied by other forms of intelligent life that there would be little hope of communication. Advocates of the Copernican principle may well be victims of an unwitting anthropocentrism when they assume mediocrity not only for humankind but for other life as well. Is it not a wonderful hubris to imagine that enough mediocrity characterizes alien life in the universe that we would actually be able to communicate with it?

"Why are there nevertheless still proponents of the SETI project?" Mayr went on to ask. "When one looks at their qualifications, one finds that they are almost exclusively astronomers, physicists, and engineers. They are simply unaware of the fact that the success of the SETI project is not a matter of physical laws and engineering capabilities but a matter of biological and sociological factors."[20]

What, we may well ask, do chemists, physicists, and astronomers know that biologists do not? Or rather, what hidden assumptions have physicists tacitly made that the biologists reject? One of my colleagues is fond of pointing out that physicists make their living by simplifying physical problems to the bare essentials, in order to cope with problem solving, whereas biologists thrive in a world of intricate and fascinating complexity. Physicists love the principles of uniformity and mediocrity, for they reduce complexity and tend toward simplicity, a process that helps them get on with their business. If friction renders the study of motion too difficult, then consider an idealized world without friction—in a first approximation our planetary system provides such a case. Isaac Newton made brilliant progress by assuming that celestial motions are the frictionless counterparts of terrestrial motion. The tool of simplicity helped reveal for the first time the underlying unity of celestial and terrestrial motions, a foundation stone in the physical uniformity of the universe.

Despite a commitment to uniformity, a physicist or chemist, seeing the beautiful efficiency and optimal design of a particular protein (such as the examples described in Michael Denton's *Nature's Destiny*), will not assume that the atoms have fallen into place randomly. Even if the scientist does not remember the absurdly low probabilities calculated half a century ago by Pierre Lecomte du Noüy (who estimated the probability that a two-thousand-atom protein could be randomly formed as something like one part in 10^{321}), the investigator will realize at once that random shakings are not the way to make a protein.[21] Rather, he will assume that catalytic processes and natural pathways aid in the building of such a complex molecule, and that the existence of such mechanisms does not violate the principle of uniformity. The physicist, more readily than the biologist, and probably unwittingly, is thus making room for design.

"Design" should not necessarily be taken to mean the detailed working out of a preordained pattern. A combination of contingency and natural selection can produce organisms exquisitely attuned to their environment, true marvels that stagger our imaginations. But contingency and natural selection do not create the extraordinary physical and chemical conditions—the solubilities, the diffusion coefficients, the bonding strengths, and so on—that permit the existence of such marvels. It is like having a giant and very complex Lego set supplied without a blueprint. There may be no architect with a plan for the final product, but there is the designer of the set of little interlocking parts. And the existence of the set itself cries out for something to be built with it.

The astronomers and physicists who assume that extraterrestrial intelligence is inevitable and ubiquitous are essentially saying that the set is rigged, that in some way it is designed not just to allow for intelligent life, but to make it likely. Paul Davies goes straight to the heart of the matter, in saying, about the idea that mind is in some sense predestined to arise in the universe: "This viewpoint, though prevalent, again conceals a huge assumption about the nature of the universe. It means accepting, in effect, that the laws of nature are rigged not only in favor of complexity, or just in favor of life, but also in favor of mind. To put it dramatically, it implies that mind is written into the laws of nature in a fundamental way."[22] He goes on to describe the search for life elsewhere in the universe as "the testing ground for two diametrically opposed world views."

The one view, that intelligent life emerges at best very rarely through extraordinary and improbable contingencies, encapsulates a strict Darwinian understanding: humankind is a glorious accident. The other view, that the universe is abundantly inhabited by intelligent creatures, carries the hidden assumption of design and purpose, in other words, of teleology.

For at least a century and a half scientists have dismissed a role for teleology—final or goal-directed causes—in science. As Ernst Mayr wrote (as a strict Darwinian), "Cosmic teleology must be rejected by science. . . . I do not think there is a modern scientist left who still believes in it."[23] Yet in their endorsement of the Copernican principle, the enthusiasts of extraterrestrial intelligence have opened a fascinating back door onto a goal-directed cosmos. In assuming the presence of other accessible alien intelligent life in the universe, they accept the existence of design principles that make life resembling our own a natural feature of the cosmos. In fact, they apply mediocrity to the aliens as well as to ourselves, to make them similar enough for possible interaction. The SETI proponents are, in effect, making an arrogant claim for our mediocrity: they use the Copernican principle to set a standard for other intelligence in the universe sufficiently similar to that for *Homo sapiens* so that we could actually hope to communicate with each other.

Atheists and theists alike may be disconcerted and challenged by the conclusion that the Copernican principle provides an opening to teleology. I am personally persuaded that a superintelligent Creator exists beyond and within the cosmos, and that the rich context of congeniality shown by our universe, permitting and encouraging the existence of self-conscious life, is part of the Creator's design and purpose. Yet like many Christians steeped in a conservative ethos that human beings are central to God's plan, my gut reaction is to disparage the possibility of the existence of intelligent life on other worlds. But I remind myself, Beware! Not only is such a view inconsistent with the notion that the universe has been deliberately established as a potential home for self-conscious contemplation, but it sets unwarranted human limitations on God's creativity.

Today, from a theological framework, the God-given

contingencies drive us toward both a humble approach and an insistent curiosity to investigate what is in the greater universe. Whether atheist or theist, a thoughtful person can only stand in awe of the way the universe seems designed as a home for humankind. In the words of an eminent living cosmologist, "Humility in the face of the persistent great unknowns is the true philosophy that modern physics has to offer."[24] We can hope that our increased scientific understanding will eventually reveal more to us about God the Creator and Sustainer of the cosmos.

Is mediocrity a good idea? It is not a fundamental principle of science, although the allied notion of the uniformity of nature is a powerful unifying principle that somehow seems essential to scientific progress; however, mediocrity as a guide to understanding our place in the universe seems to me a generally unexamined ideology, and not one to which I would readily subscribe. We human beings are the most extraordinary creatures we know about, and part of our glory is that we can imagine we are *not* the most remarkable creatures in the entire universe. But as the physicist John Wheeler once suggested to me, perhaps the universe is like a large plant whose ultimate purpose is to produce one small exquisite flower. Perhaps we are that one small flower. Quite possibly mediocrity is *not* a good idea!

For me the universe is a more coherent and congenial place if I assume that it embodies purpose and intention, and that leads me to another question: Dare a scientist believe in design? That is the question I will examine next.

A fragment of the Allende meteorite, which fell in Mexico in 1969.
The dark fusion crust formed in its flight through the earth's
atmosphere, having been fractured, reveals the tiny white
chondrules dated as 4.6 billion years old.

2

Dare a Scientist
Believe in Design?

We shall look at a particular example and shall conclude
that, when it comes to complexity and beauty of design,
Paley hardly even began to state the case.

—Richard Dawkins, *The Blind Watchmaker* (1987)

*L*et me begin by turning the clock back a little more
than five hundred years, to 7 November 1492. Just before noon a brilliant fireball exploded over Switzerland,
and near the Alsatian village of Ensisheim a stony meteorite plunged three feet into the ground. Three weeks
later Emperor Maximilian rode into town and, puzzled
by the stone, consulted his counselors. They decided exactly what it was: a miracle, a wonder of God, a sign of
favor toward the emperor, who ordered that the stone be
displayed at the local church.

In the five centuries since 1492, people have commonly come to accept meteorites, comets, the northern lights, and other such mysteries as natural, normal phenomena and not as miraculous signs from God. Many highly intelligent people nevertheless somehow wish they were still in a world where brilliant fireballs are generally seen as miraculous events. That is to say, they long for a universe where God continually and dramatically intervenes in the natural world.

In 2004 the Gallup poll asked:

Which of the following statements comes closest to your views on the origin and development of human beings?

1. Human beings have developed over millions of years from less advanced forms of life, but God guided this process.

2. Human beings have developed over millions of years from less advanced forms of life, but God had no part in this process.

3. God created human beings pretty much in their present form at one time within the last ten thousand years or so.

Of the respondents, 45 percent chose the third statement, that God created human beings within the last ten thousand years or so. This option apparently calls for the mysterious creation of *Homo sapiens* out of the dust of the earth. Clearly a belief in the miraculous is pervasive in our culture today. In scientific explanations, however, miracles are barred. Although science cannot rule out miracles, they play no role in acceptable explanations.

This was not always the case. When Isaac Newton, during that astonishing eighteen-month period of intense creativity that led to his *Principia,* realized that gravitation was universal, that every mass attracted every other mass in the universe, he noticed that the planets would not entirely fall under the rule of the sun but would attract one another as well. What could keep such a complex system in adjustment? Newton decided that God's continuing care would be required to prevent the planets from running wild. This came in for discussion in the famous correspondence between Newton's surrogate, Samuel Clarke, and the German philosopher Leibniz. Leibniz replied that it was a mean notion of the wisdom and power of God which would imply He could not have gotten the universe right in the first place. Leibniz added, "I hold that when God works miracles, he does not do it in order to supply the wants of nature, but those of grace."[1]

Since then, following Leibniz, scientists have gradually excluded God's miraculous interventions from their explanations of natural events, though miracles continue to stand just outside the gates of scientific inquiry. The current brouhaha over Intelligent Design is ultimately an argument over the role of the miraculous, though it is seldom discussed in such stark terms. Since I believe in a created universe, one fashioned with staggering intricacy and beauty, am I an advocate of Intelligent Design? Dare I, as a scientist, believe in design? And is there a difference between believing in design and accepting Intelligent Design, capital *I* and capital *D?* These are the issues before us.

One of the most impressive and remarkable developments in astronomy in the past half century is our appreciation of the fact that the universe has a history that has unfolded over some billions of years. It would lead me too far afield to explicate in detail the several lines of evidence that convince scientists of the antiquity of our cosmic environment, but let me begin with the Allende meteorite, which exploded over the Mexican village of Pueblito de Allende in 1969. When the meteorite was formed out of the primitive material of the solar system, it contained minuscule amounts of radioactive elements, which, with their very long half-lives, decayed slowly. Analysis of the ratio of lead isotopes remaining in the meteorite, as well as of the ratio of argon isotopes resulting from the gradual decay of radioactive potassium, allows the object to be dated quite precisely: it has an age of 4.6 billion years. The Allende meteorite is now the oldest known macroscopic object on earth. The sample shown in the frontispiece to this chapter is one fragment from nearly a ton of material that has been recovered. These fragments date back to the birth of the solar system itself, whereas the rocks on the surface of the earth have been continually reprocessed and therefore do not preserve a record of their original antiquity.

The ages determined for Allende and several other meteorite falls accord closely with the age of the sun itself, which has been established through detailed computer models of the sun's internal structure and the rate at which it fuses hydrogen atoms into helium, its basic source of energy. These calculations indicate that the sun has shone for roughly five billion years, and that it has spent about half its nuclear fuel.

Our observed universe is about three times older than the Allende meteorite. From the headlong rush of distant galaxies, we can calculate back to a beginning, to the Big Bang itself, which is now dated to 13.7 billion years before the present. An independent method, which yields a comparable result, comes from the evolution of globular clusters, gravitationally stable balls of tens of thousands of stars found in a halo around our Milky Way Galaxy. As is the case with the sun, the age is derived from complex and sophisticated computer modeling of how they consume their nuclear fuel. The most luminous and profligate stars of the original configurations have long since expended their nuclear resources, and the observed absence of such stars allows us to date the clusters.

With respect to the Big Bang, some astronomers have resisted the notion that the universe had a beginning, a singularity that smacks all too much in their philosophy of a miraculous event. Many were delighted, however, by the congruence between the words of Genesis 1, "Let there be light!" and a hypothesized Big Bang in which the universe began with a mighty burst of energetic photons. Not just this room, or the earth, or the solar system, but the entire visible universe was squeezed into a dense dot of pure energy. And then came the explosion. "There is no way to express that explosion," writes the poet Robinson Jeffers;

> . . . All that exists
> Roars into flame, the tortured fragments rush away
> from each other into all the sky, new universes

Jewel the black breast of night; and far off the outer
    nebulae like charging spearmen again
Invade emptiness.[2]

I can recall vividly, from the time I was a young
postdoc, the point when astronomers began to appreci-
ate one of the most astonishing features of this cosmic
event, the incredible balance between the outward energy of expansion and the gravitational forces trying to
pull everything back together again. Because in the expansion itself any slight imbalance in either direction
would be hugely magnified, the initial balance had to be
accurate to about one part in 10^{59}—a ratio of 1 to 1-fol-
lowed-by-fifty-nine-zeros, an unimaginably large number. Had the original energy of the Big Bang explosion
been less, the universe would have fallen back in on itself long before there was time to build the elements required for life and to produce from them intelligent,
sentient beings. Had the energy been greater, it is quite
likely that the density, and hence the gravitational pull,
of matter would have diminished too swiftly for stars
and galaxies to form. The balance between the energy of
expansion and the braking power of gravitation had to
be extraordinarily exact—to such a degree that it seems
as if the universe must have been expressly designed for
humankind. This is the classic example of what astro-

physicists and cosmologists refer to as fine-tuning, and at that point the universe was fine-tuned indeed. If you are looking for design, how about this? Surely a beneficent Creator was at work to produce a universe fit for intelligent life!

But some nagging problems remained in this reconstruction. Astrophysicists could calculate back in time, closer and closer to the beginning—the perceived universe becoming denser and denser and hotter and hotter as they ran their calculations backward—until finally, when they reached a point about 10^{-43} second from the zero point, their physics broke down because of the uncertainty principle. By then, at a second split so fine that no clock could measure it, our entire visible universe was compressed into a tiny ball of pure energy that could pass through the eye of a needle. And that situation brought up a very interesting puzzle. Time as we know it belongs to our universe. Before the beginning there was no time. There was no "when" then. Why did all of the minisphere explode at the same time? In 10^{-43} second light cannot travel very far—not even across an atom, much less across the eye of a needle. How could a signal from one side trigger a staggeringly creative event in synchronism on the other side, where time may not yet have begun?

An intriguing solution to this conundrum was suggested three decades ago, named by Alan Guth, one of its inventors, the inflation scenario. This split-second event must be distinguished from the expansion of the universe initiated by the Big Bang explosion, an expansion that is still ongoing in our universe. In contrast, inflation was over and gone in a moment, a subtheme superimposed on the Big Bang itself, but a brief event whose consequences mold the universe as we find it today. In the unimaginably compressed and hot state of the universe immediately after its inception, the basic forces of the universe would be merged into one. Then, in the inflationary subtheme to the Big Bang, at a very early instant, just as the gravitational force separated out from the other basic forces of the universe, for a moment gravitation acted as a force of repulsion, and space expanded by billions of billions of billions of times—actually even more. That is to say, the entire part of the universe now visible to us was once exceedingly tiny, so tiny that in the opening moments all the parts could be connected by impulses moving at the speed of light.

Now, one consequence of the inflation scenario is that it automatically brought into balance gravitational braking and the energy of expansion. Thus, today we would say that this incredibly fine-tuned balance, to one part in 10^{59}, was no accident; or, one might argue, direct evidence of God's designing hand in the opening moments of the universe. This is not what Aristotle might have called a fact-in-itself, but rather a "reasoned fact"— a phenomenon with an explanation; for the balance between the energy of expansion and the gravitational braking must inevitably follow, as a consequence of that split second of extraordinary inflation. Has this evidence for God's being and presence in our universe gone away? A "fact-in-itself" might call for God's fine-tuning hand in the history of the universe, whereas the march of science has obviated the necessity of such a role. But a "reasoned fact" still resonates with the magnificence of God's original designs that have made such a universe possible. Here we have the difference between the occurrence of contingent miraculous events and fundamental planning for the nature of the universe itself.

And our universe does seem singularly congenial as a home for intelligent life. L. J. Henderson's *Fitness of the Environment* extolled the remarkable life-enhancing properties of water, as well as pointing out the unique properties of the carbon atom, including the fact that carbon can bond with itself in a vastly larger number of combinations than any other atom.[3] It is this wonderful property that makes complex organic chemistry possible.

Of course, these unique properties would have been of little avail in fostering life, had it not been for the substantial abundance of oxygen and carbon. But since hydrogen and oxygen rank first and third, respectively, in cosmic abundance, water is guaranteed to be present throughout the universe, and carbon comes in fourth in order of cosmic abundance. If we are allowed to think of God in anthropomorphic terms, we would have to say, "Good planning!" Curiously enough, neither oxygen nor carbon emerged in the first three minutes of the Big Bang. At first glance, this might be labeled God's Goof. That's how the physicist George Gamow felt when he discovered the flaw in the nature of the light elements that prevented the heavier elements from forming. In the first minute of the Big Bang, energetic photons were transformed into protons, which fused into deuterium (nuclear particles of mass two), tritium (nuclear particles of mass three), and alpha particles (which would serve as mass-four nuclei of helium atoms). But there was no stable mass five, so at that point the fusion process stopped, well short of the twelve needed for carbon or the sixteen for oxygen. Gamow, who had an impish wit, then wrote his own version of Genesis 1:

> In the beginning God created Radiation and Ylem (a mixture of protons and neutrons). And the Ylem was without shape or number, and the nucleons were rushing madly upon the face of the deep.
>
> And God said: "Let there be mass two." And there

was mass two. And God saw deuterium, and it was good.

And God said: "Let there be mass three." And there was mass three. And God saw tritium, and it was good.

And God continued to call numbers until He came to the transuranium elements. But when He looked back on his work, He saw that it was not good. In the excitement of counting, He had missed calling for mass five, and so, naturally, no heavier elements could have been formed.

God was very disappointed by that slip and wanted to contract the universe again and start everything from the beginning. But that would be much too simple. Instead, being Almighty, God decided to make heavy elements in the most impossible way.

And so God said: "Let there be Hoyle." And there was Hoyle. And God saw Hoyle and told him to make heavy elements in any way he pleased.

And so Hoyle decided to make heavy elements in stars, and to spread them around by means of supernova explosions. But in doing so, Hoyle had to follow the blueprint of abundances which God prepared earlier when He had planned to make the elements from Ylem.

Thus, with the help of God, Hoyle made the heavy

elements in stars, but it was so complicated that neither Hoyle, nor God, nor anybody else can now figure out exactly how it was done.[4]

But far from being a design flaw in our universe, the absence of mass five seems essential to our existence. The lack of a stable mass five means that the element-building in the stars takes place as a two-step process: first, hydrogen is converted into helium in the hot nuclear cauldrons at the cores of stars; and then, once helium is abundant, it is built up into heavier atoms, in a second process. Because helium has a mass of four units, the fusion of two or three or four helium nuclei results in atoms of mass eight or twelve (carbon) or sixteen (oxygen), thus skipping over the unstable mass five. This second process requires a much higher temperature in the stellar interiors, one that is not reached until much of the hydrogen fuel has been exhausted—in the case of a star like the sun, only after about ten billion years. This guarantees a long, steady lifetime for sunlike stars. It is of course this tedious process that provides the stable solar environment in which the evolutionary biological sequences can work themselves out.

If mass five were not absent, that could not happen. Suppose that mass five were stable. Then, in the opening minutes of the universe, characterized by the overwhelming abundance of protons (each with a mass of one unit), atom-building could have taken place as mass increased by steps of one, right up the nuclear ladder toward iron. This would have left no special abundance of carbon (mass twelve) or oxygen (mass sixteen), two essential building blocks of life. The actual process, which uses helium nuclei, preferentially builds the life-giving abundance of carbon and oxygen. Fortunately, in many stars much more massive than the sun the exhaustion of hydrogen took place considerably more rapidly, and then, in their death throes, these massive stars dispersed their ashes into space. The sun and its solar system formed in a later generation, enriched with these heavier elements. You and I and the Allende meteorite are made of star stuff, recycled cosmic wastes. And that, incidentally, is why a habitable environment arises only in a very old universe, one in which there has been time to build the critical elements.

What at first glance appeared to be God's mistake turns out to be one of the Creator's most ingenious triumphs. Certainly, the way our universe works, the fact that it takes a very long time to generate the heavier elements depends critically on the lack of a stable mass five. In the absence of a nuclear ladder with easy steps (each adding one in mass), the ladder goes up by steps of four. Ancillary processes are required to fill in the less abundant intermediate elements, such as nitrogen or sodium. Thus, the production of the various heavier elements is a complicated matter. For example, when the element beryllium (mass eight) is formed by the fusion of two helium nuclei, it tends to fall apart quickly, an evanescence that would leave an inadequate base for the next step up to carbon (mass twelve).

The late Fred Hoyle played a leading role in figuring out these processes, as Gamow implies in his parody of Genesis. Basing his hunch on the evidence that carbon-dependent life does exist, Hoyle predicted that there must be something special about the carbon nucleus to enable it to overcome the comparative instability of the mass-eight beryllium, something that would account for the high abundance of carbon in the universe. His prediction was borne out and helped bring the Nobel Prize for Physics to Willy Fowler, whose experimental work showed that a special state in the carbon nucleus did indeed exist. I am told that Fred Hoyle said that nothing shook his atheism as much as this discovery.

These are only a few of the remarkable details of the physical world that make intelligent life on earth possible. The Astronomer Royal, Sir Martin Rees, has written a book entitled *Just Six Numbers,* in which he describes six physical numbers that, if changed slightly, would produce a cosmos in which life could not exist.

These include such things as the ratio of electrostatic to gravitational attraction. The coincidences seem too great to be ignored or written off as accidental. Somehow, the universe seems rigged, "a put-up job," as Fred Hoyle expressed it. One possibility is that the universe was intentionally and intelligently designed. As Aristotle might say, this could be a final cause, a meta-explanation that transcends ordinary scientific explanations, just as metaphysics transcends physics.

Or perhaps for reasons still opaque to us, all of these constants are forced to have these values because there is no alternative. Perhaps the great holy grail of physics, that "theory of everything," will show why they must be this way, just as surely as two plus two must equal 4, and thus show that God had no choice. Or perhaps a multitude of choices and a multitude of other universes exist, forever unobservable by us, in their own space-time arenas in places totally disconnected from our own. This scenario could conceivably yield an unimaginable, infinite number of unobservable parallel universes. I am tempted to say that this is sheer metaphysics, whereas Sir Martin, who has endorsed this multiverse view, assures me that it is not. Given myriad universes, with a variety of properties, then naturally the intelligent, self-contemplating *Homo sapiens* would be found in precisely the universe where the roulette brought up the numbers conducive to complex life forms! Frankly, I find this an unconvincing solution. The mere fact that one rare universe is just right seems miracle enough. And by the way, if there is in the end only one way to make a universe, all the multiverses would have to be identical. Their rationale would evaporate, and we would be left with just one astonishingly congenial universe.

Impressive as the evidence for design in the astrophysical world may be, however, I personally find even more remarkable the indications from the biological realm. As Walt Whitman proclaimed, "A leaf of grass is no less than the journey work of the stars."[5] I would go still further and assert that stellar evolution is child's play by comparison with the complexity of DNA in grass or mice. Whitman goes on, musing:

> The tree toad is a chef-d'oeuvre for the highest,
> And the running blackberry would adorn the parlors of heaven,
> And the narrowest hinge in my hand puts to scorn all machinery,
> And the cow crunching with depress'd head surpasses any statue,
> And a mouse is miracle enough to stagger sextillions of infidels.

Even Hoyle, normally agnostic, was staggered by these apparent touches of design, writing, "A common sense interpretation of the facts suggests that a super-intellect has monkeyed with physics, as well as with chemistry and biology, and that there are no blind forces worth speaking about in nature." In his allusion to the biology, he in effect agreed that the formation of, say, DNA is so improbable as to require a designing principle. Such biochemical arguments were popularized about sixty years ago by Pierre Lecomte du Noüy in his book *Human Destiny*. Lecomte du Noüy wrote: "Events which, even when we admit very numerous experiments, reactions, or shakings per second, *need an infinitely longer time than the estimated duration of the earth in order to have one chance, on an average, to manifest themselves can, it would seem, be considered as impossible in the human sense.*"[6]

Lecomte du Noüy went on to say, "To study the most interesting phenomena, namely Life and eventually Man, we are, therefore, forced to call on anti-chance, as Eddington called it; a 'cheater' who systematically violates the laws of large numbers, the statistical laws which deny any individuality to the particles considered."[7]

The game plan for evolutionary theory, however, is to find the accidental, contingent ways in which these unlikely and seemingly impossible events could have taken place. The evolutionists seek not an automatic schema, but some random pathways that could be at least partially retraced from the fragmentary historical record by induction.

Ever since Charles Darwin published *On the Origin of Species* in 1859, his evolutionary theory has engendered controversy. From the very beginning some churchmen found its insights quite acceptable, whereas other believed it was incompatible with their understanding of the place of humankind in the scheme of things.[8] Similarly, some atheists opposed it, while others found it entirely agreeable. Then, as now, therefore, the reaction to evolution could not be said to cleave entirely along religious lines. The controversy reached a dramatic point with the Scopes trial of 1925, which was organized by the businessmen of Dayton, Tennessee, to bring a share of publicity to their town.[9] Because evolution remains a hot-button issue even today, I cannot address the question "Dare a scientist believe in design?" without touching on it.

My late colleague Stephen Jay Gould liked to remark that "evolution is a fact, not a hypothesis." Now, if he meant that rock layers can be identified in sequence and dated with ages going back hundreds of millions of years, and that these rock layers show a steady progression of life forms from the simple to the increasingly complex, then this is as close to a scientific fact as we can get today. But linked to these observations is a stunning hypothesis: All living creatures are formed by *common descent with modification*. This means that every living creature has a parent. Eels do not just slither out of the mud with no parents, and flies are not generated from garbage. It also means that not every child is the same species as its parent, something that offends scriptural literalists because Genesis 1:24 says, "And God said, Let the earth bring forth the living creature after his kind, and it was so."

Today, of course, even high school students study a great deal more about genetics than Darwin ever knew, and about its molecular basis. We now know how many chromosomes are in each human cell, something that was still in doubt when I was a graduate student, and we know that the chromosomes are miniature coils of DNA coding for the many proteins that constitute the molecular machinery that runs the cells. We recognize that a pattern of nucleic acid triplets provides the information about which amino acid is linked next in the sequence that makes the proteins. And we understand that a mutation can alter this pattern and thereby produce a slightly different protein, which can have interesting consequences for the organism. In fact, in our bodies mutations are taking place all the time, but usually nothing comes of them. Only when mutations take place in germ cells is there any chance that they will be transmitted.

Let me digress for a moment to describe a particular case that has been intriguing me. Among the highly inbred population of Amish in Lancaster County, Pennsylvania, a rare pathology occasionally occurs, known as six-finger dwarfism.[10] There are approximately seventy-five known cases, in about half of which the children were stillborn, and the remarkable circumstance is that in every instance *both* parents could trace their ancestry to a single Amish couple who immigrated from Europe to Pennsylvania around 1750. Now, in our cells the chromosomes come in pairs, so if something goes wrong with one, the pairing chromosome can furnish valid genetic coding, while the other remains recessive. One partner in that original Amish couple carried a defective gene, but it was not expressed, because it was recessive. The spouses thus had no clue about the insidious trait they were transmitting to their descendants.

On average, half their offspring carried the mutant recessive gene, which was passed on to their descendants in turn. Because of the intermarriage of the descendants, a child had a small chance of getting the defective genes from both the mother and the father, and in that case the dreadful consequences manifested themselves. A few years ago the single altered nucleic acid in the DNA coding was discovered.[11] Just one substitution in the blueprint for making a critical enzyme is enough to cause a host of attendant changes, including the development of a sixth finger and a crippled spinal column. Obviously, the mutation must trigger some control mechanism in the complex genetic system, so that the alteration in one nucleic acid sets off an entire cascade of further missteps in reading off the DNA instructions. The case shows how intricate and extraordinary the control mechanisms are in the DNA coding, and how cautious we have to be in making claims about macroevolutionary changes and the ways they can or cannot occur.

I am certainly not claiming that these six-fingered children are of a different species from their parents, but this example offers at least some idea of how a succession of mutations could produce a new species. And everyone will agree that on the basis of merely random mutations the process is extremely unlikely to come up with successful products. The anthropologist Irven DeVore likes to compare the prospect of producing favorable mutations with trying to tune your MG by standing fifty paces away and firing at it with a shotgun. One pellet might accidentally hit the valve just right to adjust the engine, but it is far more likely that the car will be destroyed before that can happen. (Of course, if you are firing at an entire parking lot full of cars, there is a finite chance that you can drive out with one intact.)

I think you will agree that most mutations will be undesirable and will therefore tend to die out, whereas beneficial mutations will tend to be transmitted to successive generations. Darwin was mightily impressed with the fecundity of nature, especially as he saw it expressed in the equatorial regions, while he was on the voyage of the *Beagle,* and he was also convinced of the antiquity of the process of natural selection. The great age of the earth was a novelty for many of his readers, so he had to dedicate a fair amount of text to the evidence for it. Granted these two fundamental conditions, fecundity and antiquity, combined with variations (that is, mutations), he then brilliantly argued that the competition inherent in natural selection would take care of the rest, even though he had essentially no information concerning how the variations themselves could arise.

Many people in our country are, of course, still ill equipped to understand the solid basis for believing in the great antiquity of the earth and are still locked into a primitive scriptural literalism that leads erroneously to a conclusion that the earth is only a few thousand years old. The leading theorists of the so-called Intelligent Design (ID) movement are *not* in that camp, although it is a knee-jerk reaction of many scientists to assume that they are simply a front for young-earth creationists. In fact, these ID theorists are much more sophisticated than that. At issue for them is whether random mutations can generate the incredible amount of information content required to produce even the simplest of cells, and whether even the great antiquity of the universe could make this possible. Here science, dealing with extremely low probabilities balanced against vast numbers of opportunities, is frankly on very shaky turf.

Before I press this point further, let me recall an occasion some years ago when I participated in a remarkable conference of theists and atheists in Dallas. One session considered the origin of life, and a group of Christian biochemists argued that the historical record is nonscientific, since it is impossible to perform scientific experiments on history. Furthermore, they amassed considerable evidence that the current scenarios concerning the chemical evolution of life were untenable.

I soon found myself in the somewhat anomalous position that to me the atheists' approach was much more interesting than the theists'. That particular group of Christian biochemists had concluded that ordinary science did not work in such a historical situation—that is, with respect to the origin of life—and they attempted to delineate an alternative "origin science" in which the explicit guiding hand of God could make possible what was otherwise beyond any probability. As I said at the outset, many among us like to think of God's miraculous intervention in natural history as a relatively frequent occurrence. The reason I admired the atheist biochemists so much was that they had not given up trying to find a natural, nonmiraculous path to the origin of life. They were still proposing ingenious avenues whereby catalytic effects in the chemistry made the events far more likely. "Let us not flee to a supernaturalistic explanation," they said; "let us not retreat from the laboratory."

It might be that the physics and chemistry of life's origins *are* forever beyond human comprehension, but I see no way to establish that scientifically. Therefore, it seems to me to be part of science to keep trying, even if ultimately no answer is accessible. Apparently, this reasoning has some cogency, because the ringleader of the theistic group, Charles Thaxton, at least partly backed off from his original position, and today we hear little about origin science. Meanwhile, a new generation has reclothed some of these ideas and presented them under the rubric "Intelligent Design." Citing some of the same bits of evidence that impress me, about the remarkable hospitality of our universe as a home for intelligent life, as well as about the improbability of the random formation of intelligent life, the Intelligent Design theorists press the case still further and argue that some evolutionary steps make sense only when taken as a large bundle, a form of macroevolution. A frequently cited example is the whiplike flagellum used for propulsion by the bacterium *E. coli*. This miniature rotator requires a series of special proteins for its construction, and, as the controversial argument runs, without the complete set, the flagellum would not rotate and hence would be useless. It would strain credulity that all the required proteins could assemble at once by blind chance, the ID theorists declare, and therefore the evolutionary process requires both a designing mind and a designer's hand—in other words, a miraculous intervention. My theological presuppositions incline me to be sympathetic to this point of view, to the idea of a God who acts in the world. I believe, with the overwhelming majority of Christians, in a universe of meaning and purpose, a universe designed to be astonishingly congenial to intelligent life. Whether we look at the nature and abundance of the atoms themselves or the remarkable ratio of electrostatic to gravitational attraction or the many other details of our physical universe, we know that without these design features we would not be here. In a word, I believe in intelligent design, lower case *i* and lower case *d*.

But I have a problem with Intelligent Design, capital

I and capital *D*. It is being sold increasingly as a political movement, as if somehow it is an alternative to Darwinian evolution. Evolution today is an unfinished theory. There are many questions about details it does not answer, but those are not grounds for dismissing it. As I said, I cited the case of six-finger dwarfism to show how intricate and extraordinary the control mechanisms are in the DNA, and how cautious we have to be in making claims about macroevolutionary changes. Presumably, both the evolutionists and the Intelligent Design theorists must rely, in their speculations, on mutations, and the divide occurs in the way they think about them. As I indicated, most mutations are disasters, but perhaps some inspired few are not.

Can mutations be inspired? Here is the ideological watershed, the division between atheistic evolution and theistic evolution; and frankly, it lies beyond the capability of science to prove the matter one way or the other. In 1998 I invited two of my most brilliant friends to participate in a debate over the question, "Is the universe designed?" One was Steven Weinberg, a Nobel physics laureate and an articulate champion of atheism. The other was John Polkinghorne, a distinguished English physicist, then the only ordained member of the Royal Society, and, by the way, a previous William Belden Noble lecturer. In his opening remarks Steve

Weinberg announced, with characteristic arrogance, that most people are not entitled to be atheists because they haven't thought enough about the matter.[12]

In listening to the debate, I was very much struck by the fact that Steve Weinberg did not offer any scientific reasons for his atheism; and that is how it should be, for "Is the universe designed?" is not a scientific question, *pace* the Intelligent Design enthusiasts. As Polkinghorne pointed out in his own remarks, we will learn the answer neither by looking for items trademarked "Heavenly Construction Company" nor by coming upon objects stamped "Blind Chance Rules."[13] The reason is simple. The question I posed is one without answer in the scientific sense. It is a metaphysical question, whose answer will come only out of metaphysical reasoning.

Science will not collapse if some practitioners are convinced that there has occasionally been creative input into the long chain of being. Are mutations blind chance, or is God's miraculous hand continually at work, disguised in the ambiguity of the uncertainty principle? Or we could be more subtle, and ask whether God designed the universe in the first place to make possible the catalysts and unknown pathways that enable the formation of life. Earlier I mentioned the incredible odds, as calculated by Lecomte du Noüy, against the chance formation of a protein molecule. Given that proteins do exist and that a mechanistic science has been highly successful, the overwhelming reaction has been simply to ignore Lecomte du Noüy, since he must be so obviously wrong. But is he? For science to overcome the odds, it is necessary for us to postulate the existence of those pathways, and it is of course precisely the challenge of science to discover such pathways. But is not the existence of such pathways also evidence of design? And are they not inevitable? These are questions that materialists do not want to hear.

Still, this does not get the Intelligent Design theorists off the hook. I think they are making a serious error of category when they propose that Intelligent Design should be taught alongside the theory of evolution in science classes. Let me show you why.

I am holding a fragment of the Allende meteorite in my hand, and I propose to let go. You will not be surprised by what happens. It drops to the floor. Why? I could say that it is God's will that the stone falls. I am not being facetious, for I firmly believe that God is both Creator and Sustainer of the universe. I could say that in every moment God re-creates the universe, and the meteorite, in each successive re-creation during its descent, is slightly closer to the ground than before. I could declare that part of God's sustaining power consists in the maintenance of the laws of the natural world. In fact, the very expression "laws of nature," from the time of Boyle and Newton, derives from the concept of divine law, and it is probably not accidental that science arose in such a philosophical/theological environment. However much we might assert that the stone fell because of divine will, though, such a statement does not pass muster as a scientific explanation. What science requires is a broader explanatory schema, one that links falling apples and meteorites with the fall of the moon, and that enables us to calculate the trajectories of the Voyager rocket blasted off to Saturn or the spin of skaters like Sarah Hughes or Michelle Kwan.

After Newton published his *Principia,* critics, especially Leibniz, complained that he had not really "explained" gravity, and that for the moon to be pulled toward the earth by invisible means was just plain occult and superstitious. Newton was sufficiently troubled by the critique that he added a "General Scholium" to the second edition of his book, in which he admitted that he could not explain the essence of gravity. "I feign no hypotheses," he wrote in a very famous statement. Yet he continued to maintain that space was the "sensorium of God" (whatever that means!) and that somehow God's sustaining action could let the moon or an apple know immediately that the earth was attracting it. Thus we see that there are multiple levels of explanation for any phenomenon. God's role as Sustainer can be described in Aristotelian terms as a final cause, the ultimate teleological reason something happens. Over the years since the Scientific Revolution, however, one vast panoramic scientific picture has been put together that is singularly successful in explaining *how* the universe works, what Aristotle would call an efficient cause. Today scientists, as scientists, play by the rules of a game of coherence, putting together an integrated picture of how things work, without recourse to the miraculous or to ultimate reasons. Essentially, scientists' quest takes place in the realm of efficient causes. Thus, much as I might believe that the universe is best understood in terms of intelligent design, I don't think that would get a spacecraft to Mars or explain how the laser in the grocery store checkout line works. As a scientist I accept methodological naturalism as a research strategy.

Many leading theorists of ID argue that the evidence for intelligent input into the evolutionary process is overwhelming. With regard to final causes, those theorists make a good case for a coherent understanding of the nature of cosmos. But they fall short in supplying any mechanisms to serve as the efficient causes that primarily engage scientists in our age. Intelligent Design does not explain the temporal or geographical distribution of species. Intelligent Design does not help us understand why Hawaii, in comparison with the older continental areas, has so few species, and why there would be flightless birds on the islands. It does not shed any light on why the DNA in yeast is so closely related to the DNA in human chromosomes. As a philosophical idea, ID is interesting, but it does not replace the scientific explanations that evolution offers. It simply does not offer any insight regarding the numerous related skeletal patterns—for example, the five bones of the coelacanth's fins and the five bones of the gorilla's hand, excellent examples of the sort of mystery illuminated by the hypothesis of common descent with modification.

There is, however, another side to the coin. Some of the most spirited and vocal defenders of evolutionary theory, such as Richard Dawkins, use their stature as scientific spokesmen as a bully pulpit for atheism. Dawkins claims that evolution makes atheism intellectually fulfilling.[14] I suppose he single-handedly makes more converts to Intelligent Design than any of the leading Intelligent Design theorists. Or we might take Cornell University's evolutionary biologist and historian of science William B. Provine, who, in defending the gospel of meaninglessness, has written that if modern evolutionary biology is true, then lofty desires for divinely inspired moral laws or some kind of ultimate meaning in life are hopeless.[15]

Evolution as a materialist philosophy is ideology, and presenting it as such essentially raises it to the rank of final cause. Evolutionists who deny cosmic teleology and who, in placing their faith in a cosmic roulette, argue for the purposelessness of the universe are not articulating scientifically established fact; they are advocating their personal metaphysical stance. This posture, I believe, is something that should be legitimately resisted. It is just as wrong to present evolution in high school classrooms as a final cause as it is to fob off Intelligent Design as a substitute for an efficacious efficient cause.

Have I seriously faced the query: Dare a scientist believe in design? There is, I shall argue, no contradiction between holding a staunch belief in supernatural design and working as a creative scientist, and perhaps no one illustrates this point better than the seventeenth-century astronomer Johannes Kepler. He was one of the most inventive astronomers of all time, a man who played a major role in bringing about the acceptance of the Copernican system through the accuracy of his tables of planetary motion.

One of the principal reasons Kepler was a Copernican arose from his deeply held belief that the sun-centered arrangement reflected the divine design of the cosmos: the sun at the center was the image of God, the outer surface of the star-studded heavenly sphere was the image of Christ, and the intermediate planetary space represented the Holy Spirit. These were not ephemeral notions of his student years, but a constant obsession that inspired and drove him through his entire life. To his teacher Michael Maestlin back in Tübingen he wrote, "For a long time I wanted to be a theologian; for a long time I was restless. Now, however, behold how through my effort God is being celebrated in astronomy!"[16]

Today Kepler is best remembered for his discovery of the elliptical form of the planets' orbits. This discovery and another, the so-called law of areas, are chronicled in his *Astronomia nova,* truly the New Astronomy. In its introduction he defended his Copernicanism starting from the conviction that the heavens declare the glory of God:

> If someone is so dense that he cannot grasp the science of astronomy, or so weak that he cannot, without offending his piety, believe Copernicus, I advise him to mind his own business, to quit this worldly pursuit, to stay at home and cultivate his own garden, and when he turns his eyes toward the visible heavens (the only way he sees them), let him with his whole heart pour forth praise and gratitude to God the Creator. Let him assure himself that he is serving God no less than the astronomer to whom God has granted

the privilege of seeing more clearly with the eyes of the mind.[17]

Kepler's life and works provide central evidence that an individual can be both a creative scientist and a believer in divine design in the universe, and that indeed the very motivation for the scientific research can stem from a desire to trace God's handiwork.

To believe in a designed universe requires accepting teleology and purpose. And if that purpose includes contemplative intelligent life that can admire the universe and can search out its secrets, then the cosmos must have properties congenial to life. For me, part of the coherency of the universe is that it is purposeful—though probably it takes the eyes of faith to accept that idea. But if a person accepts that understanding, the principle that states that our universe must be well suited to life also becomes the evidence of design. This brings to mind a few lines in Whitman's *Leaves of Grass:*

A child said *What is the grass?* fetching it to me with full
    hands;
How could I answer the child? I do not know what it is
    any more than he. . . .
I guess it is the handkerchief of the Lord,
A scented gift and remembrancer designedly dropt,

Bearing the owner's name someway in the corners, that
we may see and remark, and say *Whose?*[18]

In reflecting on the question of design, I have attempted
to delineate a subtle place for it in the world of science.
Intimations of design can offer persuasion regarding the
role of divine creativity in the universe, but not proof.
Science remains a neutral way of explaining things, not
anti-God or atheistic. Many people are extremely un-
comfortable with a way of looking at the universe that
does not explicitly presuppose the hand of God. En-
dorsing the scientific viewpoint, as I remarked at the
outset, does not mean that the universe is actually godless, just that science generally has no other way of
working.

Even in the hands of secular philosophers, however,
the modern mythologies of the heavens, the beginnings
and endings implied in the Big Bang, give hints of ultimate realities beyond the universe itself. Milton Munitz,
in his closely argued book *Cosmic Understanding,*[19] declares that our cosmology leads logically to the idea of a
transcendence situated beyond time and space, giving
the lie to the notion that the cosmos is all there is or was
or ever will be. Munitz, in arriving at the concept of
transcendence, describes it as unknowable, and that is
somewhat paradoxical, for if the transcendent is un-

knowable, then we cannot know that it is unknowable. Could the unknowable have revealed itself? That the unknowable might have communicated with us defies logic, but it does not contradict coherence. For me, it makes sense to suppose that the transcendence, the ground of being, in Paul Tillich's formulation, the serendipitous creativity of Gordon Kaufman's *In Face of Mystery*, has revealed itself through prophets in all ages, and supremely in the life of Jesus Christ.

It was Galileo who wrote that the reality of the world was dually expressed in the Book of Scripture and in Nature, and these two great books could not contradict each other, because God was the author of both.[20] So, just as I believe that the Book of Scripture illumines the pathway to God, I also believe that the Book of Nature, in all its astonishing detail—the blade of grass, the missing mass five, or the incredible intricacy of DNA—suggests a God of purpose and a God of design. And I think my belief makes me no less a scientist.

A cube of uranium 238, part of the German atomic bomb
project of World War II. Shown here actual size.
Harvard University collection.

3

QUESTIONS WITHOUT ANSWERS

Two things fill my mind with ever new and increasing wonder and awe, the more often and persistently I reflect upon them: the starry heaven above me and the moral law within me.

Epitaph of Immanuel Kant, Königsberg

From the treasures of Troy to the plunder amassed by the conquistadors of Peru to William Jennings Bryan's 1896 "Cross of Gold" speech or even to gold-plated satellites shot into space, the fascination with gold is deeply embedded in our global culture. Yet amazingly, all the gold mined in historical times could just about fit into a cube fifty-four feet on a side.[1] The weight of such a cube would be roughly equivalent to the weight of the steel produced in four hours in the United States. How can we account for this vast discrepancy between the abundance of iron and the rarity of gold, or is this a question without an answer?

Apart from a few capped teeth, I carry no gold with me, but I have brought a small cube of uranium (shown in the frontispiece to this chapter). Like gold, uranium is a dense element, and that small cube weighs about five pounds. If it were gold, it would be worth about $30,000, which is one reason I brought uranium instead of gold! Curiously, in the cosmos as a whole gold is at least ten times more common than uranium, but here at the surface of the earth uranium is about five hundred times more abundant than gold. It is not too difficult to understand these differences. Gold, like other heavy elements, such as platinum or iridium, tends over long ages to sink down into the core of the earth. Uranium, too, is a heavy element, but unlike those other precious elements, it is radioactive, and the heat it generates warms the rocks, so that a convection pattern develops. The ever-so-slow motion of the rocks carries the uranium back up to the surface of the earth. This convection circulation is extremely important in the history of our planet, for over hundreds of millions of years it helps build the continental zones, gives rise to continental drift, and buries the carbon-dioxide-bearing rocks.

It is fascinating to notice that the so-called half-life of

uranium, a measure of how rapidly the element decays radioactively, is 4.5 billion years, comparable to the age of the earth itself. If the half-life were much longer, uranium would not produce nearly as much of the heat that stirs the mantle of the earth; if it were much shorter, most of the uranium would be gone by now; and in either case the surface of the earth would be much different. Is this another example of fine-tuning, part of the set of anthropic physical constants that make intelligent life possible on earth? That is a tricky question to answer, so I simply draw your attention to it as something to ponder.[2]

Speaking of questions, if you had asked a physicist or an astronomer back in 1930 about *why* iron is so much more common than gold or uranium, he might well have told you to get lost. To ask the reason for the disparity in abundance between iron on the one hand and gold and uranium on the other was then definitely to pose a question without an answer—even a question for which no answer was foreseeable. But science marched on, and within three decades it became a perfectly sensible question and astronomers could begin to outline how elements form in stars, how iron represents an end point in the evolution of normal stars, and how the precious heavy elements in particular result from rare supernova explosions that generate these atoms in a swift fiery shower of neutrons and then spew them out into space. We are all definitely star stuff, but gold and uranium are the stuff of supernovas.

Scientists are fond of asking questions, particularly those with answers. Last summer *Science* magazine celebrated its 125th anniversary by asking 125 questions, such as What is the universe made of? Are we alone in the universe? and What genetic changes have made us uniquely human? These are not easy questions, and they may not have easy answers, but they are the sort of questions that science is good at eventually answering. In fact, it was an essay by my sometime coauthor Alan Lightman, who pointed out that science owes its great success to choosing questions that can be answered, that inspired me to address questions *without* answers.[3] Lightman allowed that questions without answers could be important, but science tries not to tangle with them. It was, I believe, Erasmus who said that these are the questions for the philosophers.

Probably the most profound of these philosophical questions is, Why is there something rather than nothing? or, Does the universe have a purpose? These are teleological questions, definitely in Aristotle's final causes department, and not for science to grapple with. Steve

Weinberg has written, "The more the universe seems comprehensible, the more it also seems pointless,"[4] but that is a personal philosophical speculation that falls well beyond the purview of science.

Einstein famously posed another conundrum in a 1936 essay where he remarked that the eternal mystery of the world is its comprehensibility. He added: "It is one of the great realizations of Immanuel Kant that the setting up of an external world would be senseless without this comprehensibility."[5] Why is it that we can make so much progress in understanding the cosmos? Why is the universe comprehensible? Here indeed is a deep mystery.

These questions arise from the wellspring of the human mind, and it is only there that the answers can be sought. Why is there something rather than nothing? In speculating about possible answers to these seemingly unanswerable questions, we require the discussion to be comprehensible and coherent, even though, from a scientific perspective, the answers may lack convincing proofs. But before I press this further, let me offer a historical example of the elusive role of truth in science.

First, here is some background regarding the heliocentric system, which Nicolaus Copernicus introduced in his *De revolutionibus,* a book published in 1543, the very year in which he died. Copernicus himself does not state directly what induced him to work out the sun-centered arrangement, apart from some rather vague dissatisfaction with the inelegance he perceived in the traditional geocentric pattern. But Copernicus was nothing if not a unifier. In Ptolemaic astronomy each planet was more or less an independent entity. The result, Copernicus wrote in the preface to his book, was like a monster composed of spare parts, a head from here, the feet from there, arms from yet another place. In Ptolemy's system the path of each planet was generated by a main circle and a subsidiary circle, the so-called epicycle. Copernicus discovered that he could eliminate one circle from each set by combining them all into a unified system, and when he did this, something almost magical happened. Mercury, the fleetest planet, circled closer to the sun than did any other planet. Lethargic Saturn automatically circled farthest from the sun, and the other planets fell into place in between, arranged in distance from the sun according to their periods of revolution.

In his monumental treatise on the heavenly revolutions, Copernicus summed up his aesthetic vision: "In no other way do we find a wonderful commensurability and a sure harmonious connection between the size of the orbit and the planet's period of revolution."[6] By commensurability, Copernicus meant that all the planetary orbits were locked into their dimensions by the "common measure" provided by the earth's orbit. Once this heliocentric unification was accomplished, the system showed other advantages. It accounted, for example, for the curious fact that whenever Mars or Jupiter or Saturn went into its so-called retrograde motion, the planet was always directly opposite the sun in the sky. In the principal cosmological chapter of his book, Copernicus noted that the heliocentric arrangement finally provided a natural explanation for this otherwise unexplained coincidence. He mentioned as well that heliocentrism explained why the retrograde motion of Jupiter traced a smaller arc than that of Mars, and why that of Saturn was still smaller. Copernicus' achievement was a vision in the mind's eye; he had no observational evidence that the earth moved. Even so, as Copernicus' only student and disciple, Georg Joachim Rheticus, put it, "All these phenomena appear to be linked most nobly together, as by a golden chain; and each of the planets, by its position and order and every inequality of its motion, bears witness that the earth moves."[7]

Yet these explanations were not enough to win the day. Astronomers of the sixteenth century belonged to a long tradition that had distinguished astronomy from physics. At the universities astronomy was taught as part of the quadrivium, the four advanced topics of the seven liberal arts. The astronomer instructed his students in the understanding of celestial circles, the geometry of planetary mechanisms, and the calculation of positions required for drawing up horoscopes. But the physical nature of the heavens was described not in Aristotle's *De coelo,* but in his *Metaphysica,* and that text was the province of the philosophy professor. The distinction was rather clearly stated in the anonymous "Introduction to the Reader," added to *De revolutionibus* by the Lutheran clergyman Andreas Osiander, who had served as proofreader for the publication. "You may be worried that all of liberal arts will be thrown into confusion by the hypotheses in this book," Osiander wrote (and I paraphrase): "Not to worry. It is the astronomer's task to make careful observations, and then form hypotheses so that the positions of the planets can be calculated for any time. But these hypotheses need not be true nor even probable. A philosopher will seek after truth, but an astronomer will just take what is simplest. And neither will find truth unless it has been divinely revealed to him."[8]

With few exceptions, astronomers of the sixteenth

century accepted Osiander's stance, taking *De revolutionibus* as a recipe book for calculating planetary positions, but not as a true physical description of the cosmos. The distinguished Danish astronomer Tycho Brahe noted, "On no point does it offend the principles of mathematics. Yet it ascribes to the earth, that hulking, lazy body, unfit for motion, a motion as fast as that of the ethereal torches, and a triple motion at that."[9] Thus, Tycho had no problem with the Copernican system as a mathematical construct, but he believed that Copernicus fell short with respect to physics. Surely, if the earth was spinning at a dizzying speed, stones thrown straight up would land far away. And if the earth was wheeling around the sun, how could it keep the moon in tow? These consequences would require a new physics, which was nowhere in sight.

But as a new century dawned, two astronomers in particular, Johannes Kepler and Galileo Galilei, began to argue for the reality of the heliocentric arrangement. Kepler sought a cosmic physics, not mere geometrical models. Meanwhile Galileo's terrestrial physics regarding the relativity of motion paved the way for understanding why we do not feel the earth's spin. It was not just the physics that caused a problem, however. Philosophers and churchmen must have felt threatened by the potential challenge to traditional sacred geography. Where would heaven and hell be located in the new picture? And was it not true that Joshua, at the Battle of Gibeon, asked God to command the sun, not the earth, to stand still? Clearly the task of reading the evidence was confused, scientifically as well as culturally.

Neither Kepler nor Galileo tells us precisely why he became a Copernican. Yet Galileo would say he could not sufficiently admire those who had embraced the heliocentric arrangement *despite* the violence it did to their own senses.[10] Kepler always justified his choice in terms of the Holy Trinity, but that could hardly have been his starting point. Surely it was the aesthetic appeal that arrested their attention, the sheer geometrical beauty of an arrangement that included the distant promise of a new physics. It was Kepler who first glimpsed this new physics when he discovered not only that Mars moved in an orbit with the sun as one focus of the ellipse but also that the earth in its orbit had the property of speeding up when it was closer to the sun. These discoveries had been made by 1605, though publication of Kepler's *Astronomia nova* was delayed until 1609.

It was then that Galileo turned his optical tube, not yet named the telescope, toward the heavens. In the following January he found the four bright satellites of Jupiter, and by April of 1610 his *Sidereus nuncius* was in print. There, he allowed himself a Copernican remark: "We have here a splendid argument for taking away the scruples of those who are so disturbed in the Copernican system by the attendance of the moon around the earth while both complete the annual orbit around the sun that they conclude this system must be overthrown as impossible. For our vision offers us four stars wandering around Jupiter while all together traverse a great circle around the sun."[11] I would suggest that this realization that the earth could likewise keep the moon in tow was absolutely central to Galileo's conversion to a strong, enthusiastic heliocentrism. As Copernicus said, it was a theory pleasing to the mind, and although Galileo was beginning to make it intellectually respectable to accept the earth's motion, he still had no physical proof for it.

In this context Cardinal Roberto Bellarmine, the leading Catholic theologian, wrote an often-quoted letter to Paolo Antonio Foscarini, a Carmelite monk from Naples who had published a tract defending the Copernican system. Bellarmine's letter, which was obviously intended as much for Galileo as for Foscarini, opened on a conciliatory note: "For to say that assuming the earth moves and the sun stands still saves all the appearances better than eccentrics and epicycles is to speak well. . . . But to affirm that the sun is *really* fixed in the center of the heavens and that the earth revolves very swiftly around the sun is a dangerous thing, not only irritating the theologians and philosophers, but injuring our holy faith and making the sacred scripture false."[12]

Bellarmine made very clear that he was unwilling in the absence of an irrefutable or apodictic proof to concede the motion of the earth, when he added: "If there were a true demonstration, then it would be necessary to be very careful in explaining Scriptures that seemed contrary, but I do not think there is any such demonstration, since none has been shown to me. To demonstrate that the appearances are saved by assuming that the sun is at the center is not the same thing as to demonstrate that *in fact* the sun is in the center and the earth in the heavens."

Bellarmine certainly understood Copernicus in the light of Osiander's anonymous introduction. In the opening lines of his letter to Foscarini Bellarmine stated, "First, I say that it appears to me that your Reverence and Signor Galilei did prudently to content yourselves with speaking hypothetically, as I have always supposed Copernicus did."[13] His letter raises a challenging inquiry for us: What would it have taken to convince Bellarmine that the Copernican system was the correct, physically real description of our universe? Most astronomy textbooks today, for example, list the Foucault pendulum as proof of the earth's rotation, and the annual stellar parallax as proof of the earth's yearly revolution around the sun. Would this evidence have converted Bellarmine to the Copernican doctrine, and if not, why not? Framing the question in these terms will enable us to distinguish between proof and persuasion, and to gain some insight into the matter of truth in science.

Suppose that the Foucault pendulum had been set in motion, with the shifting orientation of its swing. What would Bellarmine have made of that? Well, why not suppose that the influences of the stars that whirl around the earth each day caused the plane of oscillation of the pendulum to rotate? This is not a frivolous way out, for it is the general relativistic explanation. And what if the annual stellar parallax had been found, that tiny wiggle perceived in the positions of stars that is correlated with the earth's yearly revolution around the sun? Well, why not give each star, circling around each year, its own tiny epicycle? I think such an explanation would naturally have occurred to Bellarmine. You may immediately think of Ockham's razor and suppose that the simpler explanation would surely prevail. But remember that Ockham's razor is not a law of physics. It is an element of rhetoric, in the tool kit of persuasion. In the absence of a new physics, adopting myriad epicycles as an explanation might have posed no obstacle, in the cause of keeping the earth safely fixed.

Why is it that we today find the so-called proofs of the earth's motion—the stellar parallax and the Foucault pendulum—so persuasive, when they could not have been guaranteed to convince Bellarmine? The answer is of course that now the requisite new physics has arrived. We are post-Newtonian, and it is in the Newtonian framework that these fundamental experiments provide persuasive evidence. In fact, the Newtonian achievement was so comprehensive and so coherent that specific proofs were not really needed. No dancing took place in the streets, then, after Jean-Bernard-Léon Foucault swung his famous pendulum in 1851, nor did grand celebrations mark the announcement in 1838 of Friedrich Bessel's successful measurement of annual stellar parallax. The Copernican system no longer needed those demonstrations to win universal acceptance.

Without the new physics, Galileo could scarcely have found a convincing apodictic proof of the earth's motion. Yet he paved the way for the acceptance of the Copernican idea by changing the very nature of science. He argued for a coherent point of view, with many persuasive pointers, and his *Dialogo sopra i due massimi sistemi del mondo* (*Dialogue on the Two Great World Systems*), while not containing much new science, nevertheless made a persuasive case for the earth as a moving planet. While it would be foolhardy to claim that he changed the nature of science single-handedly, he surely played a pivotal role in the process. Today science marches on not so much via proofs as through the persuasive coherency of the picture it presents. What passes for truth in science is a comprehensive pattern of interconnected answers to questions posed to nature—explanations of *how* things work (efficient causes), though not necessarily *why* they work (final causes).

Just as we find scientific explanations credible because they hang together in a finely textured tapestry of connections, a coherency if you will, so we also expect that teleological and theological explanations will have a convincing consistency. This does not mean apodictic proofs. Too much that is eminently important in our world, like love or music or art, the whole aesthetic experience, does not depend on proof, or even on coherency. Splendid as it would be to have an irrefutable proof for the existence of a Creator, that would give us neither freedom nor choice. So that is not the way things work, like a flash of lightning or sign in the sky. "And behold, the Lord passed by, and a great and strong wind rent the mountains, and broke in pieces the rocks before the Lord; but the Lord was not in the wind; and after the wind, an earthquake; but the Lord was not in the earthquake; and after the earthquake a fire; but the Lord was not in the fire; and after the fire, a still small voice."[14]

It is that still, small voice, arising from a sense of awe and wonder and reverence, that can point us toward some tentative insights into these questions without answers. Without quite knowing what the purpose of the universe is, we can at least conjecture that somehow we are part of that purpose, and that perhaps understanding the universe is a part of that purpose. In that case, the universe might just be comprehensible because it is part of its purpose to be so. This, I would argue, is the route toward understanding such deep mysteries; and rather than believe that the universe is simply meaningless, a macabre joke, I would prefer to accept a universe created with intention and purpose by a loving God, and perhaps created with just enough freedom that conscience and responsibility are part of the mix. They may even be part of the reason that pain and suffering are also present in a world with its own peculiar integrity. This, for me, is God's universe.

Clearly, we live in a universe with a history, a very

long history, and things are being worked out over unimaginably long ages. We live in an incredibly vast cosmos, something that goes hand in hand with a long history. Stars and galaxies have formed, and elements come forth from the great stellar cauldrons. Like the little bear's porridge, the elements are just right, the environment is fit for life, and slowly life forms have populated the earth. As Darwin wrote at the end of *On the Origin of Species,* "There is a grandeur in this view of life, with its several powers, having been originally breathed by the Creator into a few forms or into one; and that, whilst this planet has gone cycling on according to the fixed law of gravity, from so simple a beginning endless forms most beautiful and most wonderful have been and are being evolved."

Are these creative forces purposeful, and in fact divine? Or does evolution imply purposelessness? We do not have to look far to find this latter view expressed, for we discover Professor E. O. Wilson writing, in the introduction to his new edition of the works of Charles Darwin: "Evolution in a Darwinian world has no goal or purpose: the exclusive driving force is random mutations sorted out by natural selection from one generation to the next. Evolution by natural selection means, finally, that the essential qualities of the human mind also evolved autonomously. The revolution begun by Darwin showed that humanity is not the center of creation, and not its purpose either."[15]

There is a curious leap of logic in Wilson's summary. It would, I think, have been difficult for Darwin to show that humanity *is* the center of creation, but probably equally difficult for him to show that humanity *is not* the center of creation. It is obviously Wilson's philosophical stance that humanity is not the center of creation, an ideology displaying a certain level of consilience, as he might put it, but no more than that.

Now, I am happy to concede that ample evidence demonstrates that natural selection is a major force at work, though I would be hesitant to say that it is the *exclusive* driving force. Natural selection provides a beautiful explanation for the proliferation of the many kinds of finches on the Galapagos, for instance, and for the ridiculously curious and imperfect adaptation there that I mentioned in the first chapter, of the red-footed booby, a duck that despite its webbed feet nests in trees on Genovesa Island. Perhaps you have had a wisdom tooth removed because you have too many teeth for the size of your mouth. That jaw is clearly not an example of intelligent design; rather, it is an imperfect adaptation that has occurred as a result of natural selection, working with the materials at hand to refashion and shorten the mammalian muzzle into a face. Darwin himself cited several examples of imperfect adaptation—though not always correctly, for his data were occasionally too sparse, but never mind. Other examples of imperfect adaptation can be readily found and explained through reference to the material that evolution had to work with.

Also part of the picture are those random mutations mentioned earlier. Let us go back to our cube of uranium and think a bit about randomness. Atoms, they are very small, to paraphrase a line from John Updike. So small that there are about six million billion billion uranium atoms in my little cube. But these atoms are unstable. Their nuclei have a tendency to eject a so-called alpha particle, thereby transmuting the uranium nucleus into another unstable nucleus, thorium 234. On average, it will take about 4.5 billion years until this ejection occurs, but many uranium nuclei will have much shorter lifetimes and others, much longer. The cube contains so many atoms, however, that statistically roughly one hundred million are decaying every second.

An unanswerable question is *when* any particular uranium atom will decay. We can establish with high statistical accuracy how many will decay per second, but we have absolutely no way to know when any particular atom will eject an alpha particle. It is as if these minute specks have free will! Within the ensemble, constraints prevail, but at the individual level "choice" exists. And since we live in a universe that is likewise indeterminate at its lowest level, the mutations caused by similarly random processes are also indeterminate. They figure among the questions without answers. Darwin himself was mystified by the source of the variations on which natural selection worked. In an 1861 letter to the eminent geologist Leonard Horner, he wrote, about the breeding of pouter pigeons: "So under nature, I believe variations arise, as we must call them in our ignorance."[16] Today we know that the variations arise at the level of the DNA coding, and that they are occurring all the time in random cells in our own bodies, but we have no more clue about why specific variations arise than we do about which of the billions of billions of uranium atoms in the cube will decay in the next minute.

Earlier I asked whether the forces shaping our universe might be divine—that is, ordained by a spirit of purpose and intention. With the openness that physics has revealed at the most fundamental levels of the universe, such forces, lying outside the gates of science, could well be present. Whether they exist might be another of those fundamental questions without answers, and as the physicist and theologian John Polkinghorne has written, "physics—or science generally—constrains metaphysics, but it does not determine it, just as the foundations of a house constrain what can be built on them, but they do not determine the actual form of the edifice."[17] We can look with awe and wonder at an unexpected mutation, regardless of whether we are religious, and the science will be the same. Let us be perfectly clear about what I am arguing. Whether the mutations are anything other than mathematically random is a question without answer *in a physical or scientific sense.* But my subjective, metaphysical view, that the universe would make more sense if a divine will operated at this level to design the universe in a purposeful way, can be neither denied nor proved by scientific means. It is a matter of belief or ideology how we choose to think about the universe, and it will make no difference how we do our science. One can *believe* that some of the evolutionary pathways are so intricate and so complex as to be hopelessly improbable by the rules of random chance, but if you do not believe in divine action, then you will simply have to say that random chance was extremely lucky, because the outcome is there to see. Either way, the scientist with theistic metaphysics will approach laboratory problems in much the same way as will his atheistic colleague across the hall. And probably both will approach some of the astonishing adaptations seen in nature with a sense of surprise, wonder, and mystery.

As Einstein said, "the most beautiful experience we can have is the mysterious. . . . Whoever does not know it and can no longer wonder, no longer marvel, is as good as dead. . . . A knowledge of the existence of something we cannot penetrate, our perceptions of the profoundest reason and the most radiant beauty . . . it is this knowledge and this emotion that constitute true religiosity."[18]

Now in titling this third chapter "Questions without Answers," I paused, because symmetry suggested that this title, like those of two preceding ones, should be framed as a question, and so I considered calling it, "Dare a Theologian Believe in Design?" In a way, the Reverend William Paley has done theology a disservice with his *Natural Theology; or, Evidences of the Existence and Attributes of the Deity.* He opens, famously, with the lines: "In crossing a heath, suppose I pitched my foot against a *stone,* and were asked how the stone came to be there; I might possibly answer, that, for anything I knew to the contrary, it had lain there forever. . . . But suppose

I had found a *watch* upon the ground, and it should be inquired how the watch happened to be in that place; I should hardly think of the answer which I had before given." I need not quote further, for you doubtless know the argument. There must be an artificer who designed the use of the watch—an invincible argument, Paley declares. To say otherwise is atheism, he says (and again I quote), "for every indication of contrivance, every manifestation of design, which existed in the watch, exists in the works of nature; with the difference, on the side of nature, of being greater and more, and that in a degree which exceeds all computation." But Paley, in linking one of his favorite words, contrivance, with design, inevitably draws a picture of a mechanician or even a tinkerer sitting there putting the biological world together bit by bit.

In more modern times we can turn to a wonderful quotation, but with some of the same problems, from the brilliant English astrophysicist and maverick cosmologist Fred Hoyle, who, in looking at some of the anthropic "evidences," wrote: "Would you not say to yourself, 'Some supercalculating intellect must have designed the properties of the carbon atom, otherwise the chance of my finding such an atom through the blind forces of nature would be utterly minuscule.' Of course you would. . . . A common sense interpretation of the facts suggests that a superintellect has monkeyed with physics, as well as with chemistry and biology, and that there are no blind forces worth speaking about in nature."[19] Paley is creating the image that God, the white-haired old man of the Sistine Chapel ceiling, gets up from his throne and goes off to his workshop to tinker some more. In Hoyle's image God goes off to his computer center and calculates. It is probably the same God who walks into the garden and says to Eve, "What is this that thou hast done?"

This is why I thought of asking, "Dare a theologian believe in design?" Systematic theologians at the Divinity School will know that God is a Spirit, that God is Light, and that God is Love—all Johannine texts[20]—and will not be confused, but amateur theologians such as myself can be easily misled, especially when we are thinking about design. The God having the creative force to make the entire observable universe in a dense dot of pure energy is incomprehensible, beyond human imagining (apart from the comparatively feeble efforts of theorists exploring the inner workings of the Big Bang). And yet we can see the consequences of this unimaginably powerful creative act: a universe congenial to the ultimate formation of life, life giving rise to intelligence that can ask questions science cannot answer. It is

God's universe. The fact that it is a universe congenial to life means that other reflective life may exist elsewhere, perhaps in abundance, and science may eventually even give a positive answer to the question whether other intelligent life exists. Of course, the answer cannot in fact be negative—it would have to remain a question without an answer, for science can never state for sure that we are the only self-conscious, thinking creatures in the universe.

I did in fact consider another question as a title for this lecture. What does it mean to be human? When *Science* magazine posed its 125 questions, the staff highlighted 25 of them, including one that is deceptively like my question, namely, What genetic changes made us uniquely human? A subtle nuance attaches here to the meaning of the word "human." It is fairly obvious that the editors of *Science,* along with many scientists, are thinking in anatomical terms, or at least in terms of those coils of DNA that make up the human genome. As I was preparing the lecture on which this chapter is based, I had an unexpected encounter with Ian Tattersall, an anthropologist whose book *Becoming Human* I greatly admire, and he remarked that anatomically modern humans arrived on the scene long before any evidence suggests they possessed those higher traits we associate with humanness. *Homo sapiens sapiens,* in sharp contrast to his contemporary Neanderthal cousins, made progressive improvements in his stone tools. And the amazing art at Lascaux and that produced even earlier in the Chauvet cave offers evidence of humans capable of abstract thought. How did these things come about? Tattersall conjectures that the invention of language, roughly two hundred thousand years ago, provides the key. It was, he suggests, as if the human brain had been prewired to take this crucial step.

In the past two decades molecular biologists have made a remarkable discovery that indicates that something else unusual took place at roughly the same time, two hundred thousand years ago. In the nuclei of our cells are twenty-three pairs of chromosomes that carry most of our genetic information, but an additional amount of DNA is found in another part of the cell, the mitochondrion. Unlike the chromosomal DNA, which is sexually shuffled at conception (since each parent contributes only one of each pair of the twenty-three chromosomes, for a full set in the offspring), so that each person's nuclear DNA is unique, the mitochondrial DNA is directly inherited from the mother, and so, apart from occasional mutations, it is the same for everyone. This sameness of the mitochondrial DNA suggests that the entire world population stems from a single source—not necessarily a lone remote mitochondrial Eve, but a comparatively small group, and it is not hard to imagine that it was the small group that invented language. With the invention of language, humanity crossed the Lamarckian divide from Darwinian evolution, in accordance with which instincts are coded into DNA, to cultural evolution, in which the human brain can begin to store more information than the chromosomes. It is this crossover that makes us distinctly human and uniquely different from the rest of the animal kingdom, the kingdom of which we are nevertheless so much a part.

Although this transformation does not precisely correspond to the Biblical story of Adam and Eve, there are interesting points of contact. Perhaps the most important verse of the first chapter of Genesis is the statement, "God created man in his own image, male and female created he them." What are these God-given attributes? I would suggest creativity, conscience, and consciousness, that is, self-consciousness, all essential human qualities. And the story of the tree of knowledge of good and evil surely describes the quintessential step toward becoming human, the origin of conscience: it was the fall into freedom, acquisition of the ability to make wrong choices and the self-consciousness to recognize wrong choices.

I deliberately chose to open this chapter by alluding

to a particular result of human ingenuity and creativity, not so much to illustrate the abundances of the elements or the indeterminism of nature, but because that particular cube of uranium is deeply symbolic of our humanness. It represents the ambiguity of human choices. It can remind us of war, of the Holocaust, of terrorism, even of nuclear suicide. It can call forth self-conscious reflection on our place in the cosmos, its past and future.

That heavy piece of metal is an incredibly historic artifact: it is one piece of the German atomic bomb project. As the European war ended in Nazi defeat, American scientists swept in to round up the German nuclear physicists, to make sure that they did not fall into Russian hands. In the process, Professor Edwin Kemble acquired this specimen and brought it back to Harvard. Local legend—mythology as it turned out—had it that this was the single specimen of pure uranium that the Germans had been able to make. Among the captured scientists was Carl Friedrich von Weizsäcker, who three decades later became a William Belden Noble lecturer. He recognized the cube when I showed it to him, but he added that the German nuclear physicists had had many of them.

Last April my wife and I went to the White Sands

Missile Range to visit the Trinity test site where the first nuclear explosion had taken place.[21] It is a bleak, seemingly God-forsaken place, but one that harbors haunting memories. In July 1945, the race with Germany was over; the German scientists had not even come close to making a bomb, but Los Alamos maintained the momentum. It was a Faustian bargain. The project leader, J. Robert Oppenheimer, later confessed, "It is my judgment in these things that when you see something that is technically sweet, you go ahead and do it and you argue what to do about it only after you have had your technical success."[22]

After the mighty explosion, brighter than a thousand suns, Oppenheimer simply said, "It worked." Later he recalled a line from the Bhagavad Gita, "I am become Death, the destroyer of worlds," and still later he said that the Los Alamos scientists had known sin. I am not making moral judgments. I want only to highlight the ambiguities of being human and the freedom that entails.

Late in the summer of 2005 reports of Hurricane Katrina filled the news. The evacuation plans seemed to have left behind the elderly, the bedridden, the lame, mothers of infants, the poor without easy means of transport. We were appalled. But why? Was this not simply Darwinian natural selection at work—the survival of the fittest? Where does altruism come in? E. O. Wilson recognized (as had Darwin before him) that altruism poses a problem for a philosophy of purposeless evolution. His sociobiology has been a valiant attempt to cope with that riddle, but altruism may well pose a question without an answer—or rather, possibly without a *scientific* answer derived from observation of the animal kingdom. It just might be that the more convincing answer lies in another arena and has to do with those God-given qualities of humanness which include conscience.

I had another chance encounter in the weeks while I was preparing the lecture on which this chapter is based—with Nancy Cartwright, a philosopher at London School of Economics who is particularly interested in the boundaries of science. She, too, was asking questions without answers, such as, "Is everything that happens in the natural world fixed by the laws of physics?" She has argued forcefully that our physical equations work wonderfully well as long as the circumstances are carefully constrained, but that it is largely an issue of faith that the equations will apply more generally. She has proposed a plausible metaphysical alternative to the tidy world ordered under the dominion of physics.[23] Hers is a world rich in different things, having different natures, behaving in different ways. Some important areas of systematic behavior are precisely predictable, but in other swaths we can predict what will happen only for the most part. She calls this a dappled world, taking the image from a 1918 poem by Gerard Manley Hopkins, which begins:

> GLORY be to God for dappled things—
> For skies of couple-colour as a brinded cow;
> For rose-moles all in stipple upon trout that swim;
> Fresh-firecoal chestnut-falls; finches' wings;
> Landscape plotted and pieced—fold, fallow, and
>   plough;
> And all trades, their gear and tackle and trim.[24]

Science, since the time of Galileo and Kepler, has been remarkably successful in describing and shaping our modern world, so successful that it cannot easily brook any doubts about its sway. Yet I am intrigued by Cartwright's exploration of the metaphysical space—the space that goes beyond *(meta* in the Greek), beyond physics. It seems to me that within the dappled universe is a theistic space, a perspective for viewing God's universe, a universe where God can play an interactive role unnoticed by science, but not excluded by science.

I would likewise subscribe to a statement that John

Polkinghorne made when he defended his belief in a designed universe in his debate with Steven Weinberg. Polkinghorne said, "I think I have good reasons for my beliefs, but I do not for a moment suppose that my atheistic friends are simply stupid not to see it my way. I do believe, however, that religious belief can explain more than unbelief can do."[25]

The questions without answers that I have addressed are, to many of my scientific friends, totally uninteresting and even irrelevant. They are quite content with thinking about questions *with* answers. But I have been thinking about questions without answers with increasing frequency over the past twenty-five years, and I believe I have caught glimpses of some of the answers, which I have shared with you. Who knows what insights I might have were I to live another quarter century? Nor can I exclude the possibility that I might have even more insights if I were resident in another place, that mysterious somewhere vaguely known as the hereafter. Whether I will enjoy that vantage point is another deep question without answer.

So to conclude these thoughts, let me turn to the prayer with which Kepler closed his *Harmonice mundi (The Harmony of the World)* of 1619: "If I have been enticed into brashness by the wonderful beauty of thy works, or if I have loved my own glory among men, while advancing in work destined for thy glory, gently and mercifully pardon me: and finally, deign graciously to cause that these demonstrations may lead to thy glory and to the salvation of souls, and nowhere be an obstacle to that. Amen."[26]

"Build thee more stately mansions, O my soul!" exclaimed
Oliver Wendell Holmes in his poem *The Chambered Nautilus*.
This sectioned specimen of *Nautilus pompilius*, from the
Marshall Islands, is in the author's collection. Holmes
used a similar image on his bookplate.

Epilogue

There is nothing I want to find out and yearn to know with greater urgency than this: Can I find God, whom I can almost grasp with my own hands in looking at the universe, also in myself?

—Johannes Kepler to an unidentified nobleman, 1613

When, in 1995, I argued for the role of a designing Creator in the universe, the philosopher Mortimer Adler pronounced mine an "excellent essay" but countered that I was skirting dangerously close to a central error in Christian apologetics.[1] All too often, he suggested, modern Christian theorists suppose that there must be nothing contingent in cosmic processes, in biology, and in evolution—that is, no role for chance. But, he argued, a predesigned universe would offer no place for freedom and choice, and he went on to attack much of modern natural theology.

Did my essay inadvertently hint that design implied

causal uniqueness? That I certainly did not intend. I replied that I meant *purpose* rather than design, *intention* but not a universe worked out in exquisite detail from a celestial blueprint. A world ordered to God's purposes could be achieved in any number of ways, not merely through a preordained design.

Surely the existence of fossils of extinct creatures shows not a universe laid out according to a plan for instant perfection, but a universe that makes itself. Most creatures that ever lived are with us no longer. Extinction is the name of the game. This recognition suggests that in some fashion the powerful transcendence that brought the universe into being, and which sustains it, has self-imposed limitations. Perhaps there is only one way to make a universe—whether that is the case is something that Einstein expressly yearned to know.[2] One of the great ideas of the Judeo-Christian tradition is the notion of God's freedom (and hence the contingency of nature), which has provided a fertile philosophical ground for the rise of modern science.[3] Since God could have made the universe in many different ways, the argument runs, it behooves the scientist to undertake experiments or observations to find out which way in fact characterizes the universe.

The idea of a vast and ancient universe making itself, which I have to some extent adopted in the foregoing ruminations, yields only a distant God of large numbers. Awesome as the creating transcendence is, it is the sort of deity that few would worship. What good is a God who does not interact with creation? This question comes not only from devout adherents of the great monotheistic religions, but from skeptical atheists as well.

The Jewish people, in their scriptures, interpreted history as the story of God's actions in the world.[4] *But where was God when the Holocaust happened?*

In January 2006, the church bells rang out joyously in Tallmansville, West Virginia, when for a few hours it seemed that most of the miners in the Sago coal mine disaster had survived. Shortly thereafter that proved not to be the case, and the anguish was palpable in the evening television news. "Where is God when we need Him? Is He really there?" asked a distraught woman. When I was seventeen, my only brother, biking at dusk on his paper route, was struck by a car, and a few hours later he breathed his last in the hospital. A quarter of a century later, in one of the last entries in his diary, my father, a man of deep faith, wrote that he could still not understand how God had allowed his young son to die.

This limited world is part of reality. Ours is a world of love and ecstatic joy, but also a world of suffering and excruciating pain. It is not a world of all or none, but a dappled world, where chance and randomness join with choice and inexorable law. Why creation is this way is perhaps the most unanswerable question of all.

Genesis 1 is a wonderful paean to divine creativity. The refrain, "And God saw that it was good," reflects the ultimate power and stability of creation. For a more profound insight into the way the world really is, though, we need to turn to the story of Job, where the writer wrestles mightily with a world that dispenses both rapture and cruelty.

In a dappled world characterized by both law and freedom, neither our will nor God's will is unlimited, but just because not everything in our world is sweetness and light, we cannot conclude either that God is absent or that God does not act in the universe. Consider the great Chicxulub impact event sixty-five million years ago, when an asteroid ten kilometers in diameter blasted open the great crater now buried beneath the Yucatan Peninsula, and, with the ensuing dusty darkness, canceled enough growing seasons to destroy the large dinosaurs. Opening up habitats for the proliferation of mammals was an essential step in the sequence that led to the eventual emergence of *Homo sapiens*. Without this ancient catastrophic event, neither you nor I would now be here. In a world of freedom this contingency could hardly have been a predestined event implanted in the Big Bang itself. If we accept that this is God's world and that our conscious, self-reflective existence is part of God's intention, then Chicxulub was part of God's action in the universe. If we reject the belief that God plays an active role in the universe, then we must bow to the incredible good luck of sixty-five million years ago, not to mention the billions of chance mutations that paved the evolutionary path out of the primordial ooze that led to vertebrates and finally to us.

God's world does not, however, readily lend itself to simple either-or situations. Suppose that the biophysical makeup of the universe includes preferred pathways for random chance to follow. In this way, although the decay of individual atoms in the uranium cube (described in the last chapter) cannot be predicted, their aggregate behavior is highly predictable. If such preferred pathways play a role in the history of life, then we would expect to discover widespread examples of convergent evolution, and this is precisely what the University of Cambridge paleontologist Simon Conway Morris is finding.[5] If God's intention is the emergence of self-reflective intelligence to contemplate the universe, then *Homo sapiens* might not be inevitable, but something similar could well emerge along the preferred pathways, whether or not a Chicxulub event occurred. Another implication of this line of speculation is that intelligent life to some degree like ourselves would be an expected feature and therefore one that occurs abundantly in the universe, a conclusion that should delight the SETI enthusiasts even if they have misgivings about the premise of divine intention.

If we regard God's world as a site of purpose and intention and accept that we, as contemplative surveyors of the universe, are included in that intention, then the vision is incomplete without a role for divine communication, a place for God both as Creator-Sustainer and as Redeemer, a powerful transcendence that not only can be a *something* but can take on the mask of a *someone;* a *which* that can connect with us as a *who,* in a profound I-Thou relation. Such communication will be best expressed through personal relationships, through wise voices and prophets in many times and places. The divine communication will carry a moral dimension, only dimly perceived in the grandeur of creation, yet present through the self-limitation of the Creator who has given both natural laws and freedom within its structure. Here, implications for human morality are discernible, for this view implies a kenotic or self-renunciatory ethic that is at odds with the "survival of the fittest" of evolutionary theory.[6] As Jesus said to Pilate, "My kingdom is not of this world; if my kingdom were of this world, then my followers would fight."[7]

Within the framework of Christianity, Jesus is the supreme example of personal communication from God. When the apostle Philip requested, "Show us the Father," Jesus responded, "Anyone who has seen me has seen the Father."[8] When Jesus, hanging on the cross and slowly suffocating, cried out, "My God, my God, why hast thou forsaken me?" the nature of God's self-limited, dappled world became excruciatingly clear. God acts within the world, but not always in the ways most obvious to our blinkered vision.

Kepler expressed his own anguish when he wondered whether he could find God within himself, even as he saw the remarkable harmony of the cosmos. Five years after he had penned his query, he ended his *Harmony of the World* triumphantly with a psalm: "Praise him, ye celestial harmonies, and thou my soul, praise the Lord thy Creator, as long as I shall live, for both those things of which we are entirely ignorant and those of which we know only a tiny part, because there is still more beyond. Amen."

I have touched on only a tiny part of what can be said about God's relation to this world and to us. There is indeed still more beyond. I have attempted a reasoned discourse, but perhaps I should conclude with the words of Blaise Pascal, from *Pensées* 16: "The heart has its reasons that reason does not know."

NOTES

Prologue and Pilgrimage

The modern adaptation of the Prologue to Geoffrey Chaucer's *Canterbury Tales* is by Owen Gingerich.

1. A. Einstein, *Ideas and Opinions* (New York, 1954), pp. 323–324.

1. Is Mediocrity a Good Idea?

1. Rienk Vermij, *The Calvinist Copernicans: The Reception of the New Astronomy in the Dutch Republic, 1575–1750* (Amsterdam, 2002).

2. John Polkinghorne, *Quarks, Chaos, and Christianity: Questions to Science and Religion* (London, 1994), p. 15.

3. Freeman Dyson, *Disturbing the Universe* (New York, 1979), p. 250.

4. Guillermo Gonzalez and Jay W. Richards, *The Privileged Planet: How Our Place in the Cosmos Is Designed for Discovery* (Washington, D.C., 2004).

5. Nicolaus Copernicus, *De revolutionibus orbium coelestium* (Nuremberg, 1543), folios 9v–10.

6. For a discussion of the error and how Kepler fixed it, see James R. Voelkel and Owen Gingerich, "Giovanni Antonio Magini's 'Keplerian' Tables of 1614 and Their Implications for the Reception of Keplerian Astronomy in the Seventeenth Century," *Journal for the History of Astronomy* 32 (2001): 237–262.

7. See Albert van Helden, *Measuring the Universe: Cosmic Dimensions from Aristarchus to Halley* (Chicago, 1985).

8. C. D. Andriesse, *Huygens: The Man Behind the Principle* (Cambridge, England, 2005), pp. 148–149.

9. See Michael Hoskin, *Stellar Astronomy* (Chalfort St. Giles, England, 1982), pp. 5–8.

10. Giuseppe Cocconi and Philip Morrison, "Searching for Interstellar Communications," *Nature* 184 (1959): 844–846.

11. Ibid., p. 846.

12. Edward Rosen, trans., *Kepler's Somnium* (Madison, Wisc., 1967).

13. John Wilkins, *A Discourse Concerning a New World and Another Planet . . . Another Habitable World in the Moone* (London, 1638). See Marjorie Hope Nicolson, *Voyages to the Moon* (New York, 1948).

14. *The Celestial Worlds Discover'd; or Conjectures Concerning the Inhabitants, Plants and Productions of the Worlds in the Planets,* (London, 1698), p. 83; see also p. 92.

15. See Stephen J. Dick, *Plurality of Worlds: The Origins of the Extraterrestrial Life Debate from Democritus to Kant* (Cambridge, England, 1982), especially pp. 28ff.

16. Malcolm A. Holiday, "Metabolic Rate and Organ Size during Growth from Infancy to Maturity," *Pediatrics* 47 (1971): 169–172.

17. L. J. Henderson, *The Fitness of the Environment: An Inquiry into the Biological Significance of the Properties of Matter* (New York, 1913); Michael J. Denton, *Nature's Destiny: How the Laws of Biology Reveal Purpose in the Universe* (New York, 1998).

18. The quotations from Stephen Jay Gould's *Life's Grandeur* (London, 1996), pp. 175, 214, and 216, are cited by Paul Davies in *The Fifth Miracle* (New York, 1999), p. 272; Gould makes essentially the same statement in *Wonderful Life* (New York, 1989), pp. 289 and 320.

19. Ernst Mayr, English translation, privately communicated, of his article "Lohnt sich die Suche nach extraterrestrischer Intelligenz," *Naturwissenschaftliche Rundschau* 7 (1992): 264–266.

20. Ibid.

21. Pierre Lecomte du Noüy, *Human Destiny* (New York, 1947), p. 35.

22. Paul Davies, *The Fifth Miracle* (New York, 1999), p. 271.

23. Ernst Mayr, "The Ideological Resistance to Darwin's Theory of Natural Selection," *Proceedings of the American Philosophical Society* 135 (1991): 131.

24. Martin Rees, quoting with approval the words of Joseph Silk, in *Before the Beginning* (New York, 1997), p. 6.

2. Dare a Scientist Believe in Design?

1. H. G. Alexander, ed., *The Leibniz-Clarke Correspondence* (Manchester, England, 1956), p. 12.

2. Robinson Jeffers, "The Great Explosion," in *The Beginning and the End and Other Poems* (New York, 1963).

3. L. J. Henderson, *The Fitness of the Environment* (New York, 1913).

4. G. Gamow, *My World Line: An Informal Biography* (New York, 1970), p. 127. Gamow speculated that this parody might account for his not having received an invitation to the 1958 Solvay Congress on cosmology.

5. Walt Whitman, *Song of Myself,* in *Leaves of Grass* (Boston, 1891–92), stanza 31.

6. Fred Hoyle, "The Universe: Past and Present Reflections," *Engineering and Science,* November 1981, pp. 8–12; Pierre Lecomte du Noüy, *Human Destiny* (New York, 1947), p. 35 (emphasis in the original).

7. Noüy, *Human Destiny,* p. 38.

8. James R. Moore, *The Post-Darwinian Controversies: A Study of the Protestant Struggle to Come to Terms with Darwin in Great Britain and America, 1870–1900* (Cambridge, England, 1979).

9. Edward J. Larson, *Summer for the Gods: The Scopes Trial and America's Continuing Debate over Science and Religion* (Cambridge, Mass., 1998).

10. Victor McKusick, ed., *Medical Genetic Studies of the Amish* (Baltimore, 1978).

11. Victor L. Ruiz-Perez et al., "Mutations in a New Gene in Ellis–van Creveld Syndrome and Weyers Acrodental Dysostosis," *Nature Genetics* 24 (2000): 283–286.

12. Steven Weinberg, "A Universe with No Designer," in *Cosmic Questions* (*Annals of the New York Academy of Sciences,* 950 [2001]): 169–174.

13. John Polkinghorne, "Understanding the Cosmos," in *Cosmic Questions* (*Annals of the New York Academy of Sciences,* 950 [2001]): 175–182, esp. 175.

14. Richard Dawkins, *The Blind Watchmaker* (New York, 1987), p. 6.

15. William B. Provine, "Response to Phillip Johnson," *First Things,* no. 6 (October 1990): 23.

16. 3 October 1595, in *Johannes Kepler Gesammelte Werke,* vol. 13 (Munich, 1945), 23:256–257; from Gerald Holton, "Johannes Kepler's Universe: Its Physics and Metaphysics," *American Journal of Physics* 24 (1956): 340–351, esp. 351.

17. Slightly abridged and modified from my translation in *Great Ideas Today 1983* (Chicago, 1983), pp. 321–322.

18. Whitman, *Song of Myself,* stanza 6.

19. Milton K. Munitz, *Cosmic Understanding: Philosophy and Science of the Universe* (Princeton, N.J., 1986).

20. Galileo contrasted the Book of Scripture with the revelations of nature in his "Letter to the Grand Duchess Christina," in Stillman Drake, *Discoveries and Opinions of Galileo* (Garden City, N.Y., 1957), especially pp. 182–187, where he quotes Tertullian: "We conclude that God is known first through Nature, and then again, more particularly, by doctrine; by Nature in His works, and by doctrine in His revealed word" (p. 183). Galileo does not here refer to the *Book* of Nature, but in a famous passage in his "Assayer," *Discoveries and Opinions,* pp. 237–238, he writes: "Philosophy is written in this grand book, the universe, which stands continually open to our gaze."

3. Questions without Answers

1. John Seabrook, "Invisible Gold," *New Yorker,* 24 April 1989, p. 74.

2. This point grew out of a discussion with Paul Davies, who had noticed the congruence between the half-life of uranium and the age of the earth.

3. Alan Lightman, "A Sense of the Mysterious," *Daedalus* 132, no. 4 (2003): 5-21.

4. Steven Weinberg, *The First Three Minutes* (New York, 1977), p. 154.

5. Max Jammer, *Einstein and Religion* (Princeton, N.J., 1999), p. 42.

6. My translation from Nicolaus Copernicus, *De revolutionibus orbium coelestium* (Nuremberg, 1543), bk. 1, chap. 10.

7. Georg Joachim Rheticus, *Narratio prima* (1540), as translated in Edward Rosen, *Three Copernican Treatises* (New York, 1971), p. 165.

8. My paraphrase of Andreas Osiander, "Ad Lectorem," which prefaces Copernicus' *De revolutionibus.*

9. J. L. E. Dreyer, ed., *Tychonis Brahe Dani opera omnia* (Copenhagen, 1913-1929), 4:156, lines 14-18.

10. Galileo Galilei, *Dialogue concerning the Two Chief World Systems,* Stillman Drake, trans. (Berkeley, Calif., 1953), p. 328.

11. Galileo Galilei, *The Starry Messenger,* as translated in Stillman Drake, *Discoveries and Opinions of Galileo* (Garden City, N.Y., 1957), p. 57.

12. Roberto Bellarmine to Antonio Foscarini, letter of 12 April 1615, in Galileo Galilei, *Opere,* ed. Antonio Favaro

("National Edition," Florence, 1890–1909, rpt. 1968), p. 12; abridged from the translation in Stillman Drake, *Discoveries and Opinions of Galileo* (Garden City, N.J., 1957), pp. 162–164.

13. Ibid., 171–172.

14. I Kings 19:11–12.

15. E. O. Wilson, *Harvard Magazine* 108, no. 2 (November–December 2005): 30.

16. Darwin to Horner, 14 February 1861, *Memoirs of Leonard Horner* (London, 1890), p. 300. For some years I had this original letter in my possession, a gift from Cecilia Payne-Gaposchkin, since presented to the American Philosophical Society in her memory.

17. John Polkinghorne, "Understanding the Cosmos," *Cosmic Questions* (*Annals of the New York Academy of Sciences* 950, [2001]): 175–182.

18. Max Jammer, *Einstein and Religion* (Princeton, 1999), p. 73.

19. Fred Hoyle, "The Universe: Past and Present Reflections," *Engineering and Science,* November 1981, pp. 8–12, esp. p. 12.

20. John 4:22, I John 1:5, and I John 4:8.

21. The Trinity test site is open to the public on the first Saturday in April and in October.

22. U.S. Atomic Energy Commission, *In the Matter of J. Robert Oppenheimer, USAEC Transcript of Hearing before Personnel Security Board* (Washington, D.C., 1954), p. 81.

23. Nancy Cartwright spoke on free will and the dominion of physics at the Amazing Light Symposium in Berkeley, California, on 8 October 2005, and these statements are mostly paraphrased from the text of her talk. See also Cartwright, *The*

Dappled World: A Study of the Boundaries of Science (Cambridge, England, 1999).

24. Gerard Manley Hopkins, "Pied Beauty," in *Poems* (London, 1918).

25. Polkinghorne, "Understanding the Cosmos," esp. pp. 177–178.

26. End of bk. 5, chap. 9 of *Harmonice mundi, Johannes Kepler Gesammelte Werke,* vol. 6 (Munich, 1940), p. 362; my translation is based on the ones by Charles Glenn Wallis in *Great Books of the Western World,* vol. 16, and Eric J. Aiton, A. M. Duncan, and J. V. Field, Memoirs of the American Philosophical Society, vol. 209 (Philadelphia, 1997).

Epilogue

1. Owen Gingerich, "Kepler's Anguish and Hawking's Queries: Reflections on Natural Theology," in *Great Ideas Today 1995,* ed. Charles Van Doren and Mortimer Adler (Chicago, 1995), 272–286; Mortimer Adler, "Natural Theology, Chance, and God," ibid., 288–301; Owen Gingerich, "Response to Mortimer Adler," ibid., 302–304.

2. Einstein remarked, "What really interests me is whether God really had any choice in the creation of the world," according to Gerald Holton, *The Advancement of Science, and Its Burdens* (Cambridge, Mass., 1998), p. 91.

3. See "Cosmology and Contingency," chap. 3 in E. L. Mascall, *Christian Theology and Natural Science* ([Hamden, Conn.], 1965).

4. G. Ernest Wright, *God Who Acts: Biblical Theology as Recital* (London, 1952).

5. Simon Conway Morris, *Life's Solution: Inevitable Humans in a Lonely Universe* (Cambridge, England, 2003).

6. See Nancey Murphy and George F. R. Ellis, *On the Moral Nature of the Universe: Theology, Cosmology, and Ethics* (Minneapolis, Minn., 1996).

7. John 18:36.

8. John 14:9.

ACKNOWLEDGMENTS

*L*et me now acknowledge with thanks some of those who have heightened my interest in the topics of these lectures. Robert Herrmann and Geoff Haines-Stiles helped set the stage for my involvement with the science-religion dialogue. A small consultation group that met annually at the Center of Theological Inquiry in Princeton for several years in the 1990s, to discuss God's action in the world, provided wonderful stimulation, particularly because of the participation of John Polkinghorne, Ernan McMullin, and Daniel Hardy.

ACKNOWLEDGMENTS

I am especially grateful to my colleagues David Pilbeam, Hilary Putnam, and George Whitesides for their critical reading of my manuscript. The ministers at the Memorial Church, Peter Gomes and Dorothy Austin, have enthusiastically encouraged this publication, as has Michael Fisher at the Harvard University Press. Susan Abel has offered perceptive editorial suggestions. And last but hardly least, my wife of fifty-one years, Miriam, has provided the solicitous care and inspiration that make projects of this sort possible. To all of these people I owe heartfelt gratitude.

INDEX

INDEX

INDEX